Vasil Vorsa

THE SPECTRA AND STRUCTURES OF SIMPLE FREE RADICALS

An Introduction to Molecular Spectroscopy

by GERHARD HERZBERG

NATIONAL RESEARCH COUNCIL, CANADA

DOVER PUBLICATIONS, INC., NEW YORK

Published in Canada by General Publishing Company, Ltd., 30 Lesmill Road, Don Mills, Toronto, Ontario.
Published in the United Kingdom by Constable and Company, Ltd., 10 Orange Street, London WC2H 7EG.

This Dover edition, first published in 1988, is an unabridged, corrected republication of the work first published by Cornell University Press, Ithaca, New York, 1971, in association with "The George Fisher Baker Non-Resident Lectureship in Chemistry at Cornell University."

Manufactured in the United States of America
Dover Publications, Inc., 31 East 2nd Street, Mineola, N.Y. 11501

Library of Congress Cataloging-in-Publication Data

Herzberg, Gerhard, 1904–
The spectra and structures of simple free radicals : an introduction to molecular spectroscopy / by Gerhard Herzberg.
p. cm.
Reprint. Originally published: Ithaca : Cornell University Press, c1971.
Bibliography: p.
Includes index.
ISBN 0-486-65821-X (pbk.)
1. Molecular spectroscopy. 2. Radicals (Chemistry)—Spectra. 3. Ions—Spectra. 4. Molecular structure. I. Title.
[QC454.M6H48 1988]
541.2′2—dc19 88-19050
CIP

PREFACE

During the last twenty years the intensive study of the spectra of free radicals by many spectroscopists has led to considerable advances in our understanding of their structures and of molecular structure in general. This topic formed the subject of my Baker lectures at Cornell University in the fall of 1968. In order to present the topic of free radical spectra it was, of course, necessary to give an introduction to the general field of molecular spectroscopy, selecting always those developments that are of special significance for the study of free radicals and using wherever possible the spectra of free radicals as examples. As a result the present printed version of the lectures includes a highly abbreviated summary of my three-volume monograph *Molecular Spectra and Molecular Structure*.[1] Thus I hope that the present book will be found useful not only by those interested in free radicals but also by those who are looking for a brief introduction to the field of molecular spectroscopy.

It is perhaps of interest to mention that my original plan, forty years ago, was to write a small book on the subject of no more than 200 pages. I was unable to prevent this original plan from leading to a three-volume work (of over 2,000 pages) but am very pleased that now, thanks to the invitation by Cornell University, the original plan has been fulfilled and a brief introduction to the field can be presented.

[1] G. Herzberg, *Molecular Spectra and Molecular Structure*, I: *Spectra of Diatomic Molecules* (2nd ed., 1950); II: *Infrared and Raman Spectra of Polyatomic Molecules* (1945); III: *Electronic Spectra and Electronic Structure of Polyatomic Molecules* (1966) (Van Nostrand Reinhold, New York).

An attempt to base this book on a tape recording of the lectures was soon abandoned. Nevertheless I hope that some of the flavour of the lecture method of presentation has been preserved. Naturally, in such a brief presentation of the field, choices had to be made and many fairly important topics of molecular spectroscopy had to be omitted. On the other hand, the last chapter on dissociation, predissociation, and recombination was, for lack of time, not included in the lectures but appeared to me of sufficient importance to be included in the printed version.

While some numerical data have been included in this book, for the bulk of the data the reader must be referred to the larger monograph.

Many of the numerous illustrations were taken from the monograph. I am greatly indebted to the Van Nostrand Reinhold Company for their permission to use these illustrations. I am also indebted to the Royal Society of London and the American Institute of Physics for permission to reproduce illustrations from their journals, and to the publishers of various other journals for permissions individually acknowledged in the legends.

The typing of various stages of the manuscript was done by Mrs. R. R. Boynton at Cornell, and by my secretary Miss M. P. Thompson and Mrs. H. S. Cuccaro at the National Research Council of Canada in Ottawa. Miss I. Dabrowski prepared the illustrations and helped in the proofreading. Dr. D. A. Ramsay and Miss B. J. McKenzie (now Mrs. B. Hough) read the entire manuscript and made numerous suggestions for improvements. Dr. K. P. Huber, Professor A. Lagerqvist, and my wife read the galley proofs very carefully and helped significantly to remove errors and inconsistencies. To all of them I am deeply indebted.

Finally, I should like to record my most sincere thanks to the members of the Chemistry Department at Cornell University who made my stay in Ithaca so interesting, pleasant, and enjoyable. Professor R. A. Plane, chairman of the department at the time these lectures were given, was especially generous in making certain that all arrangements for me were made in the most thoughtful manner. I am also indebted to the staff of Cornell

University Press for the care with which they carried out the preparation of this volume and for their willingness to consider all my wishes with regard to the printing of it.

G. HERZBERG

Ottawa
November 1970

CONTENTS

THE SPECTRA AND STRUCTURES OF SIMPLE FREE RADICALS

An Introduction to Molecular Spectroscopy

I. INTRODUCTION

The concept of a radical in chemistry is a very old one; it goes back to Liebig. Quoting an old text book of organic chemistry,[1] "Radicals are groups of atoms that play the part of elements, may combine with these and with one another and may be transferred by exchange from one compound into another." *Free* radicals first came to be considered after Gomberg (46)[2] at the turn of the century observed triphenylmethyl to be a chemically stable system. However, simpler radicals like CH_3, CH_2, CH are extremely short-lived species, difficult to produce and study in the free state. They are *chemically unstable* even though in general they are *physically stable;* that is, if undisturbed by collisions they do not spontaneously decompose: they have a nonzero dissociation energy.

According to the quantum theory of valence a group of atoms (a radical) when split off a parent molecule often has one or more unpaired electrons—that is, has nonzero spin (S). This circumstance has led many authors, particularly organic chemists,[3] to define a free radical as a system with a nonzero spin. Such a definition is particularly convenient for those working in the field of electron-spin resonance, since it implies that all systems and only systems that can be investigated by electron-spin resonance are free radicals. While such a definition is extremely simple and straightforward, it does have two drawbacks: according to it certain chemically stable molecules such as O_2, NO, NO_2, ClO_2 must be considered as free radicals, while on the other hand quite a number of systems that are highly reactive and short-lived,

[1] A. Bernthsen, *Text Book of Organic Chemistry*, 1st ed. 1887, 16th ed. 1924 (Vieweg, Braunschweig); English ed. 1922 (Blackie, London).

[2] Numbers in parentheses refer to the bibliography p. 209.

[3] I am indebted to Prof. J. Meinwald for a discussion of the definition of free radicals in organic chemistry.

such as C_2, C_3, CH_2, CHF, CF_2, HNO, . . . , in their singlet states ($S = 0$) are not considered to be free radicals. Indeed, one and the same system, such as CH_2, would or would not be a free radical depending on the electronic state in which it happened to be. This drawback is significant, since for a number of free radical-like systems there are both a low-lying singlet and triplet state (for example, C_2 and CH_2), and it is somewhat accidental which of the two is the actual ground state.

Therefore many physical chemists and chemical physicists use a somewhat looser definition of free radicals: they consider any *transient species* (atom, molecule, or ion) a free radical—that is, any species that has a short lifetime in the gaseous phase under ordinary laboratory conditions. This definition excludes O_2, NO, . . . but includes C_2, CH_2, CHF, . . . even in singlet states. It also includes atomic and molecular ions. In these lectures we shall adopt this somewhat loose definition of free radicals, since our aim is to discuss the spectra and structures of short-lived (transient) species. While most of the free radicals that we shall be discussing have lifetimes of less than a millisecond, we must realize that there is no sharp boundary; indeed, some of the radicals we are including have lifetimes of about 0.1 sec.

Since most simple free radicals have a very short lifetime (they are chemically unstable even though physically stable), it was only recently that attempts to isolate them and to study their structure were successful. Spectroscopy has played an important role in this development, and conversely the study of the spectra of free radicals, both diatomic and polyatomic, has contributed greatly to the understanding of molecular structure generally.

Although many atoms must be considered as free radicals, we shall not include atomic spectra in our discussions.[4] However, historically it is interesting that high concentrations of free H atoms were first obtained in 1921 by Wood (142) in a special discharge tube filled with hydrogen. The discharge showed al-

[4] See my book on *Atomic Spectra and Atomic Structure* (Dover, New York, 1944), later referred to as *AA*. More recent and more detailed books are Kuhn's *Atomic Spectra* (Longmans, London, 1970; 2nd edition) and Shore and Menzel's *Principles of Atomic Spectra* (Wiley, New York, 1968).

most exclusively the Balmer series of the H atom. Wood's tube is still used as the best source of free H atoms for studies of their reactions.

At about the same time (1920–25) certain emission bands found in flames and electric discharges were first recognized to be due to the free CH, OH, and CN radicals. The first chemical detection of polyatomic free radicals—CH_3, C_2H_5, and others—was made in 1929 by Paneth and Hofeditz (107). They produced them by thermal decomposition of metal alkyls [e.g. $Pb(CH_3)_4$] which they passed at low pressure through a heated quartz tube; the free radicals (e.g. CH_3) formed by the thermal decomposition were detected by their reaction with metals (e.g. Pb), which were placed downstream from the heated part in the form of a mirror; the reaction led to the removal of these mirrors by re-formation of the metal alkyl, thus indicating the presence of the free radicals. From the streaming velocity a lifetime of about 1 msec was obtained for these free radicals. It was not until thirty years later that spectra of CH_3 and C_2H_5 were observed.

In recent years investigations of electron-spin resonance have made many contributions to the understanding of the behaviour and structure of free radicals. Most of this work refers to the liquid and solid phase. Only quite recently have studies of gas-phase electron-spin resonance been made [Radford (115)(116), Carrington and collaborators (15)(14)(14a)]. In these lectures we shall not discuss electron-spin resonance.

Methods of observing free-radical spectra. The oldest method of obtaining the spectra of free radicals is by producing them *in emission*. Flames are the most obvious source of such emission spectra. The ordinary Bunsen burner shows in its spectrum a number of diatomic free radicals, such as CH, C_2, and OH. In a hydrocarbon flame, in addition, an extensive band system, called the hydrocarbon flame bands, appears near 2800 Å. This band system is believed to be due to the HCO free radical, but only very recently have attempts to analyze this spectrum been even partially successful. Another type of flame for the production of free radicals is the *atomic flame*, in which atomic hydrogen or oxygen or nitrogen interacts with molecules to give light emission which is due to free radicals. For example, atomic hydrogen plus NO gives a flame whose spectrum is mainly due to HNO. Active nitrogen (that is, atomic nitrogen) when acting on almost any gaseous substance will lead to the emission of some free-radical spectra. One interesting example is the flame produced when BCl_3 vapor is added to an active nitrogen stream. A strong discrete band spectrum is emitted, which, because of the condition of its production, was thought to be due to the BN molecule, until Mulliken (99) by an investigation of the isotope effect showed conclusively that the spectrum is due to BO rather than BN, the oxygen being present as an impurity.

A second method of producing emission spectra of free radicals is by way of *electric discharges*. For example, in an electric arc between carbon electrodes, the spectra of CN and C_2 are very prominent, as illustrated in Fig. 1. Other diatomic radicals may

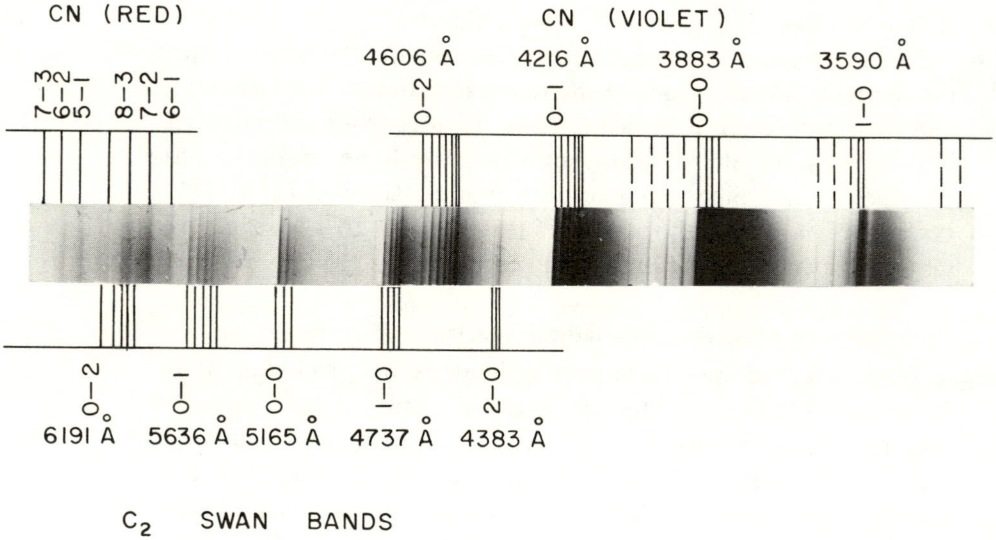

Fig. 1. Spectrum of the carbon arc in air: bands of the radicals CN and C_2 (from *MM* I, p. 31).

The bands whose leading lines are connected to the same horizontal line belong to one band system. The sequence structure of each band system (see p. 67) is clearly visible. The broken leading lines for some of the violet CN bands refer to the so-called tail bands (see *MM* I, p. 161).

be observed in such an arc if appropriate elements are present in small amounts—for example, BO in an arc fed with any boron compound. Uncondensed electric discharges through gases at reduced pressure in suitable glass tubes (Geissler tubes) show many spectra of diatomic radicals, such as NH, OH, and CN. Condensed discharges—discharges of a condenser through a discharge tube—show spectra of C_2, CH, He_2 as well as of the ions N_2^+, CO^+, CO_2^+, and many others.

A third method of producing free-radical spectra is in *fluorescence* by excitation of stable parent compounds with ultraviolet light. Such work was first done by Terenin (103)(130) and later by Style (126)(39) and their students. Irradiating suitable compounds with light in the vacuum ultraviolet, they obtained, in fluorescence, spectra of CH, NH, OH, CN, NH_2, NCO, and others. More recently Davis and Okabe (27a) and Judge and his collaborators (81) have considerably improved the fluorescence method, and have recorded spectra of CN, CH, and C_2 in the fluorescence produced by far-ultraviolet light in HCN, CH_4, C_2H_2, and other parent molecules at low pressure.

A fourth way of observing emission spectra of free radicals is by studying the *spectra of comets*. Cometary spectra are almost entirely free-radical spectra. In them the diatomic radicals CN, C_2, CH, NH, and OH as well as the molecular ions N_2^+, CO^+, and CH^+ have been found, and in addition the triatomic radicals NH_2 and particularly C_3. Clearly, these radicals are produced in comets by the absorption of far-ultraviolet solar radiation by certain parent compounds. Subsequently their fluorescence is excited by sunlight of longer wavelengths.

The production of free-radical spectra *in absorption* is in many ways more desirable but often more difficult than in emission. If a free-radical spectrum is observed in absorption, one can usually (although not always) be certain that the lower state of the observed transition is the ground state of the radical. However, it is important to realize that in order to study free-radical spectra in absorption one must have high resolution, at least if the spectrum is discrete and has sharp lines, because it is necessary to separate the background continuum on the two sides of each absorption line.

In absorption, free-radical spectra may be obtained *in flames or in high-temperature gases*. Bonhoeffer and Reichardt (11) in 1928 were the first to obtain in this way an absorption spectrum of a free radical, OH, in the laboratory by studying the absorption spectrum of water vapour at high temperature. In equilibrium a certain amount of free OH is present if the temperature is sufficiently high. Later other diatomic radicals such as CN and C_2 were observed in a similar way. The atmospheres of the sun and low-temperature stars show absorption spectra of several diatomic free radicals. Very few polyatomic free radicals have been observed either in the laboratory at high temperature or in stellar atmospheres; two that have been observed under such conditions are C_3 and SiC_2.

A number of free-radical spectra have been studied in absorption in *electric discharges*. The first was Oldenberg's (106) observation in 1934 of OH in a discharge through moist hydrogen. In a discharge through a mixture of fluorocarbons Barrow and his students (85) obtained in 1950 the first polyatomic free-radical spectrum in absorption, that of CF_2. More recently the flash-discharge method has been developed in our laboratory to study spectra of free radicals in absorption; it is illustrated by Fig. 2. A flash discharge is sent through the absorption tube F and a second flash is produced in a discharge tube S, which serves to produce the continuous background in the absorption experiment. The time interval between the two flashes can be varied in order to obtain the absorption spectrum when the concentration of the desired free radical in the tube F is greatest. The flash discharge, because it delivers a high current into a large volume of the parent compound, gives a much higher instantaneous concentration of free radicals (including molecular ions) than an ordinary discharge. In this way several free radicals and a few molecular ions have been studied.

An important modification of the discharge method is the *afterglow method*, illustrated by Fig. 3. Here a continuous discharge is produced in a side tube of the absorption tube; a mixture of the parent compound to be investigated and an inert gas is passed through this discharge and then through the ab-

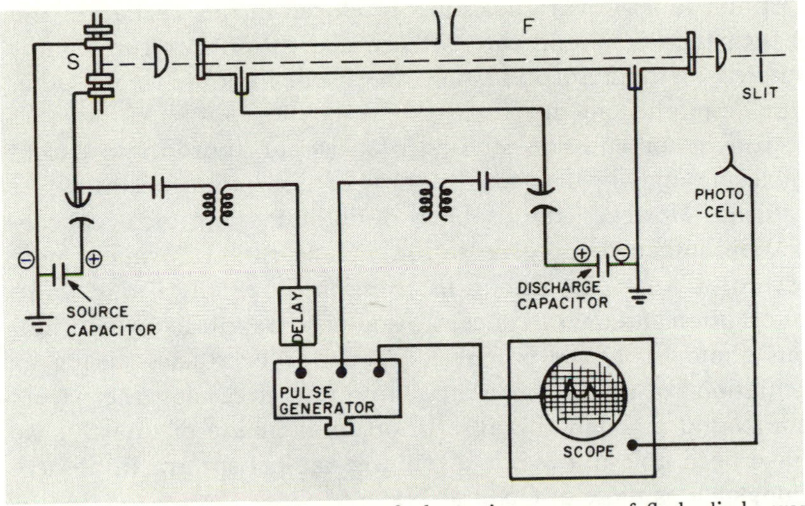

Fig. 2. Apparatus for the study of absorption spectra of flash discharges [from (66)]. [Reproduced by permission of the National Research Council of Canada from the *Canadian Journal of Physics.*]

The main flash is produced directly in the absorption tube F. The background continuum is supplied by the discharge tube S, which is fired shortly after the main flash by means of a delayed pulse from a pulse generator, which also supplies the pulses that trigger the main flash and the oscilloscope.

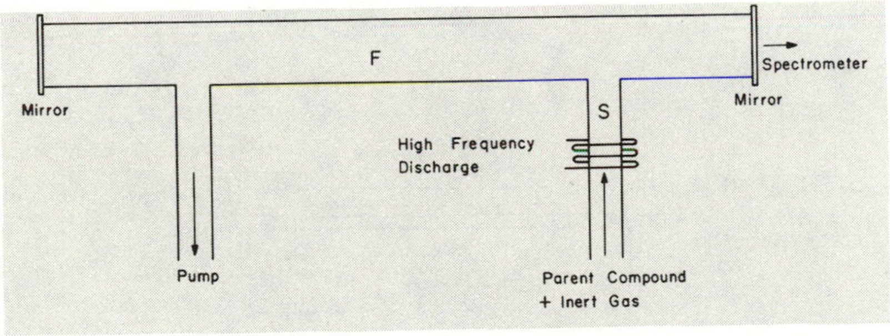

Fig. 3. Apparatus for the study of absorption spectra in the "afterglow" of a discharge.

The discharge in the side tube S is produced by a coil connected to a high-frequency generator. The parent compound (mixed with an inert gas) is decomposed in the discharge, and the decomposition products frequently give rise to an afterglow in the tube F. The absorption spectrum of the decomposition products is studied by sending light from a background source through the tube F to the spectrograph.

sorption tube. Free radicals are produced in the discharge, and if their lifetime is long enough their absorption spectrum can be studied in the absorption tube free from interference by other discharge phenomena.

Just as for emission studies, *photolysis* of appropriate parent molecules may be used for the study of absorption spectra of free radicals. However, the problem of obtaining a radical concentration sufficient for observation in absorption is much more severe than for observation in emission, especially if one wants to obtain a stationary concentration of free radicals by continuous photolysis of the parent compound. Nevertheless, using irradiation by a strong mercury lamp and streaming the parent compound through the absorption tube (made of quartz), we have been able to observe a few free-radical spectra in absorption, such as those of CN and NH_2.

Much more successful than continuous photolysis has been the *flash-photolysis* method first employed by Norrish and Porter (104) and independently developed in our laboratory. The experimental setup is shown in Fig. 4. The parent compound is

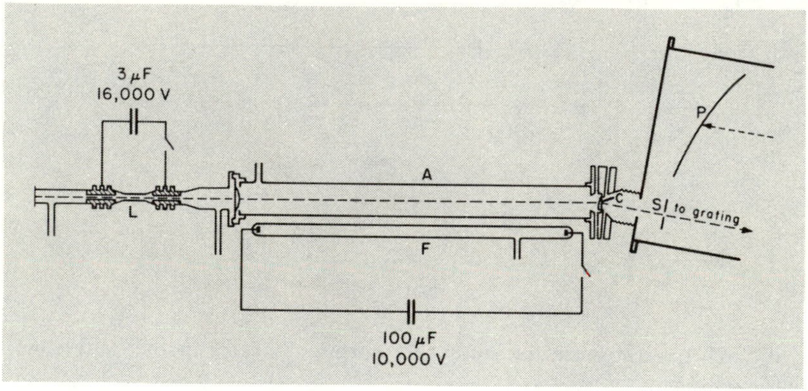

Fig. 4. Apparatus for the study of absorption spectra of free radicals in the vacuum ultraviolet by flash photolysis.

The absorption tube A is irradiated by the flash tube F. The flash tube L gives the continuous background and is fired immediately after F (at variable time intervals). At the right, part of the vacuum spectrograph is sketched (S = slit, P = plate holder).

contained in the quartz absorption tube A and is irradiated by a very strong light flash from the discharge tube F. This flash is produced by discharging a condenser battery of about 100 μF charged to 10,000 V. A second flash tube L is operated within a variable time interval (5 to 2000 μsec) after the beginning of the main flash and serves as the continuous background for the absorption experiment. Figure 4 is drawn for observations of the spectrum with a vacuum spectrograph; that is, the whole light beam is in vacuum. An LiF prism C with one cylindrical face is located between the absorption tube and the slit S and is used as a predisperser in order to separate the various orders of the grating. In this way we have been able to study absorption spectra of free radicals not only in the visible and near-ultraviolet regions, but also in the vacuum ultraviolet, where many important free-radical spectra occur.

In the visible and near ultraviolet regions we have increased the intensity of absorption by letting the light from the source flash go through the absorption tube several times. For this purpose a mirror arrangement first proposed by White (138) and slightly modified by us (9) has been used; it is shown in Fig. 5.

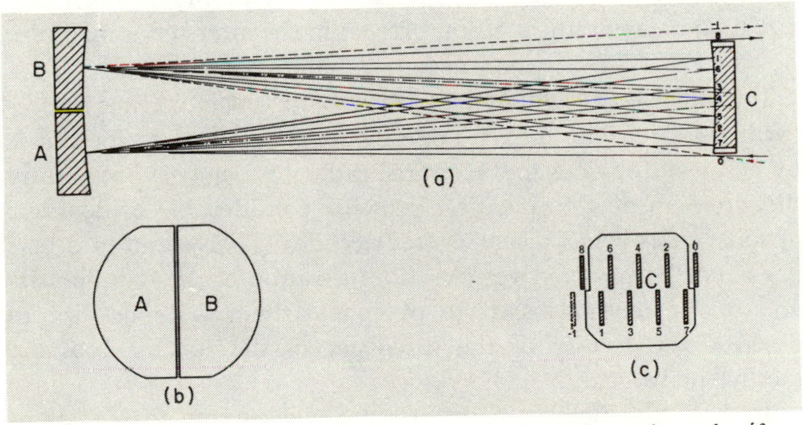

Fig. 5. Mirror system for multiple traversals of an absorption tube [from (9)].

(a) shows the path of the central beam for sixteen traversals, (b) and (c) show the shapes of the three mirrors as well as the arrangement of the slit images on mirror C.

There are three concave mirrors A, B, and C, all with a radius of curvature equal to the distance of the pair A, B from C. The light enters through a slit at 0, which is imaged by mirror A at 1. This image is imaged by mirror B at 2, and so on. The mirror C images A on B and B on A so that no light is lost except by reflection losses. Figure 5(c) shows the slit images on C for 16 traversals through the absorption tube. This number can be easily changed by rotating the mirror A by small amounts. In this way up to 100 traversals can be fairly readily obtained. The application of this method has been extremely important for the study of very weak free-radical spectra.

Finally I should like to mention still another method of obtaining free-radical spectra in absorption: the study of the spectra of distant stars. These spectra show features that can be definitely ascribed to absorption in the *interstellar medium*. In addition to a number of free atoms, the radicals CH, CH^+, CN, and OH have been unambiguously identified in the interstellar medium. Their concentration is of course extremely small, of the order of one molecule per cubic meter. A number of additional features observed in interstellar absorption have resisted all attempts at identification, but they are, at least in my opinion, very likely due to some free radical or ion present in the interstellar medium.

The problem of identification: CH_2 as an example. Once a spectrum has been obtained by one of the methods described, it is by no means obvious to which free radical it belongs. Frequently the problem of identification presents considerable difficulties: in some cases it has taken several decades to solve and in others it has not been solved yet. As an illustration of these difficulties and of the methods used to overcome them I should like to describe the history of the discovery of the spectrum of the methylene radical.

In 1941 the Belgian astronomer Swings wrote to me about a problem that had arisen in the interpretation of the spectra of comets. Figure 6 shows two spectra of a comet. In these spectra the emission bands of CN, C_2, CH can be clearly seen, but in addition there is a group of bands near 4050 Å whose

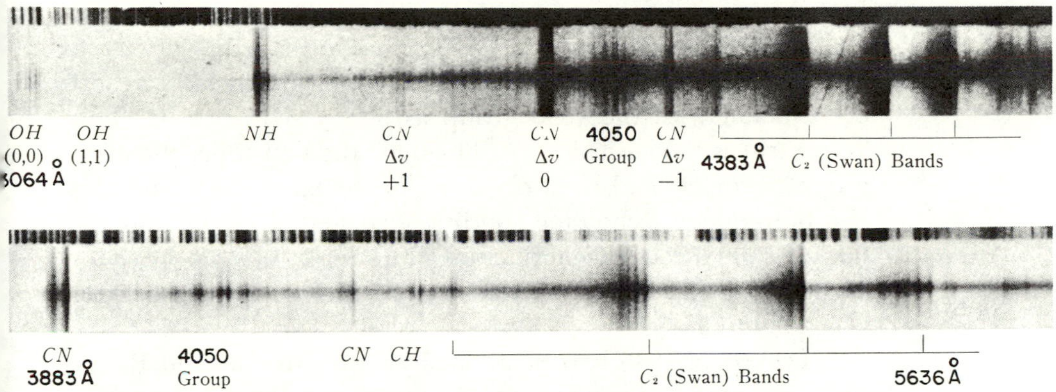

Fig. 6. Spectra of comet Cunningham (1940c); after Swings (127a). [Reproduced by permission from *Publications of the Astronomical Society of the Pacific.*]

The two spectra refer to the same comet and were taken a few days apart. The upper spectrum was taken with the quartz spectrograph, the lower one with the glass spectrograph of the McDonald Observatory.

origin nobody had been able to identify. On the basis of the structure of this spectrum, I thought I could eliminate the possibility that it was due to a diatomic free radical. Rather, the 4050 group appeared to me like a perpendicular band of a nearly symmetric top molecule (see p. 182f.), and because of the wide spacing of the subbands I concluded that it must be due to a nonlinear molecule XH_2 with a bond angle of the order of 140° (53). The most likely identification appeared to be CH_2, particularly since at that time Mulliken (100) had just predicted a spectrum of CH_2 to occur in the region 4000–4500 Å. Since CH was known to be present in comets, the identification of the 4050 group as due to CH_2 seemed eminently reasonable.

On the basis of these considerations I proceeded to do some laboratory experiments. I tried the obvious, passing a discharge through methane (CH_4) in the hope of obtaining in this discharge a spectrum of CH_2. While the continuous discharge through methane showed only well-known features such as CH and H_2, I noticed that the colour of the discharge in the first instant after it was turned on was slightly different from the later colour.

I therefore took a spectrum with the discharge turned on and off repeatedly. On this spectrum, in addition to the bands of CH, a new feature appeared precisely at 4050 Å which agreed in almost every detail with the 4050 group as observed in comets. This agreement is shown in Fig. 7. Thus, for the first time, the 4050 group of comets had been observed in the laboratory, and this had been done by choosing conditions suggested by the assumption that the spectrum was due to CH_2. Therefore, it was perhaps excusable that I felt confirmed in my belief that this spectrum was due to CH_2 (54).

However, in 1949 two Belgian physicists, Monfils and Rosen (98), repeated our experiment but replaced the hydrogen by deuterium. The spectrum that they observed was identical in every detail with the spectrum that I had observed, whereas of course small isotope shifts would have been expected had this

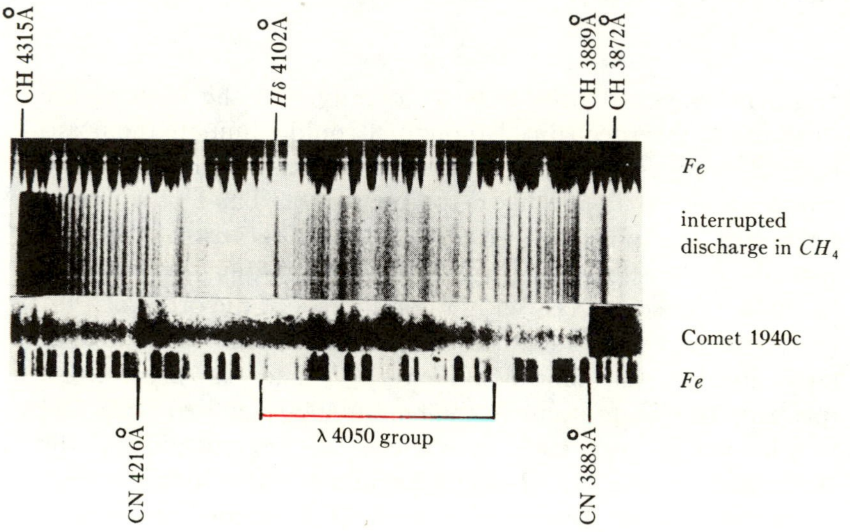

Fig. 7. The λ4050 group in the laboratory and in a comet [from (54)]. [Reproduced by permission of the University of Chicago Press from the *Astrophysical Journal.*]

The cometary spectrum is taken from the same original as Fig. 6; the laboratory spectrum is that of an interrupted discharge in CH_4. In addition to the λ4050 group the laboratory spectrum shows two well-known CH bands and the H_δ line.

spectrum been due to CH_2. Dr. Douglas and I at Ottawa immediately repeated this experiment using much higher resolution and confirmed the result of the Belgian physicists, thus establishing without doubt that neither the cometary spectrum nor the laboratory spectrum was due to CH_2. Douglas then proceeded to find the true carrier of this spectrum by using methane with C^{13} in it. He observed that the main emission band at 4050 Å in a 50-50 mixture of $C^{12}H_4$ and $C^{13}H_4$ was replaced by six bands, showing immediately that the molecule responsible for this spectrum must have three carbon atoms in it. Further consideration of the fine structure of this band left no doubt that the spectrum must be due to the free C_3 radical. At the time when Dr. Douglas established this conclusion, the C_3 radical had not even been postulated in any chemical reaction, but since that time it has been found to be one of the dominant constituents of carbon vapour as obtained by the evaporation of graphite.

The question now arose, if the 4050 group is not due to CH_2, where is the true spectrum of CH_2—or does such a spectrum not exist? It was almost ten years after the identification of the 4050 group before a spectrum of CH_2 was found. On the basis of photochemical evidence, it was well known that there are two molecules that on photolysis give CH_2—namely, ketene (CH_2CO) and diazomethane (CH_2N_2). Since the latter compound is rather explosive, we began by studying the continuous photolysis of ketene and, when that failed, by turning to flash photolysis, which had in the meantime been developed in our laboratory. Even though we extended our search into the vacuum ultraviolet, we did not find a spectrum of CH_2. As a last resort we decided that we should try diazomethane in spite of its hazardous properties. Almost the first absorption spectrum of flash-photolyzed diazomethane showed a new transient feature (of a lifetime of about 10 μsec), which turned out to be the spectrum of CH_2. We did not get this particular feature in the flash photolysis of ketene because ketene itself absorbs strongly at the same wavelengths.

We were fortunate in being able to obtain immediately, with the help of Dr. Leitch at Ottawa, a quantity of deuterated diazomethane, and in this way were able to verify, as shown by Fig. 8, that the new feature, at 1415 Å, actually does shift when hydrogen is substituted by deuterium. Thus, at least we were sure that the molecule or radical responsible for this feature contained hydrogen, but of course this observation did not yet prove that the radical was CH_2. One way of proving that *two* hydrogen atoms are present in the molecule was to take a spectrum of diazomethane that was only half deuterated; this would contain CH_2N_2, $CHDN_2$, and CD_2N_2 and therefore on photolysis, if our spectrum was really due to CH_2, it should give three bands instead of one. Actually, at first we saw only two, which was disappointing, since it seemed to indicate that only one hydrogen atom was present. However, we persisted and took pictures at higher resolution (in the fourth order of our three-meter grating

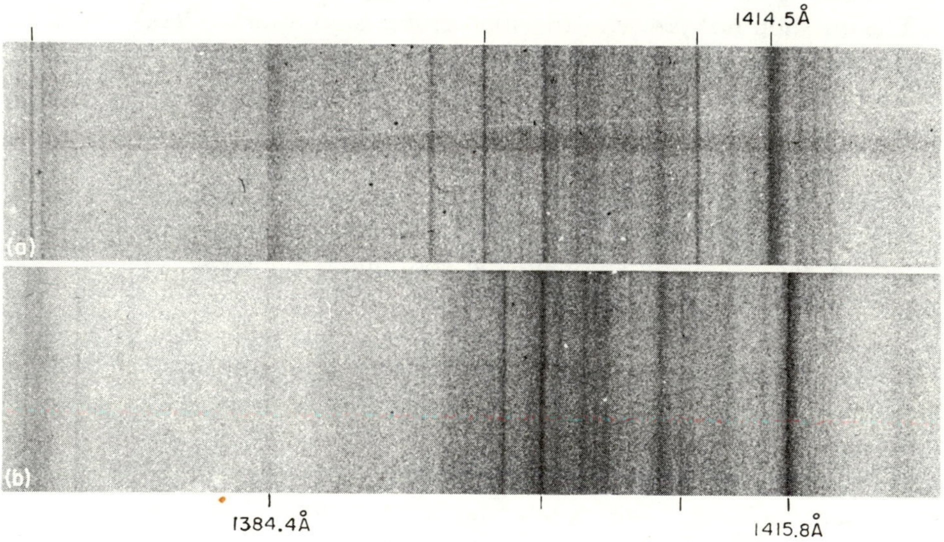

Fig. 8. Absorption spectra of flash-photolyzed (a) diazomethane and (b) deuterated diazomethane [from (57)].

The absorption band at 1414.5 Å obtained in ordinary diazomethane during flash photolysis is clearly shifted in deuterated diazomethane, as are a number of other features.

vacuum spectrograph). These high-resolution spectra showed that there are indeed three isotopic bands, proving that two H atoms are present.

These spectra, reproduced in Fig. 9, also show for two of the

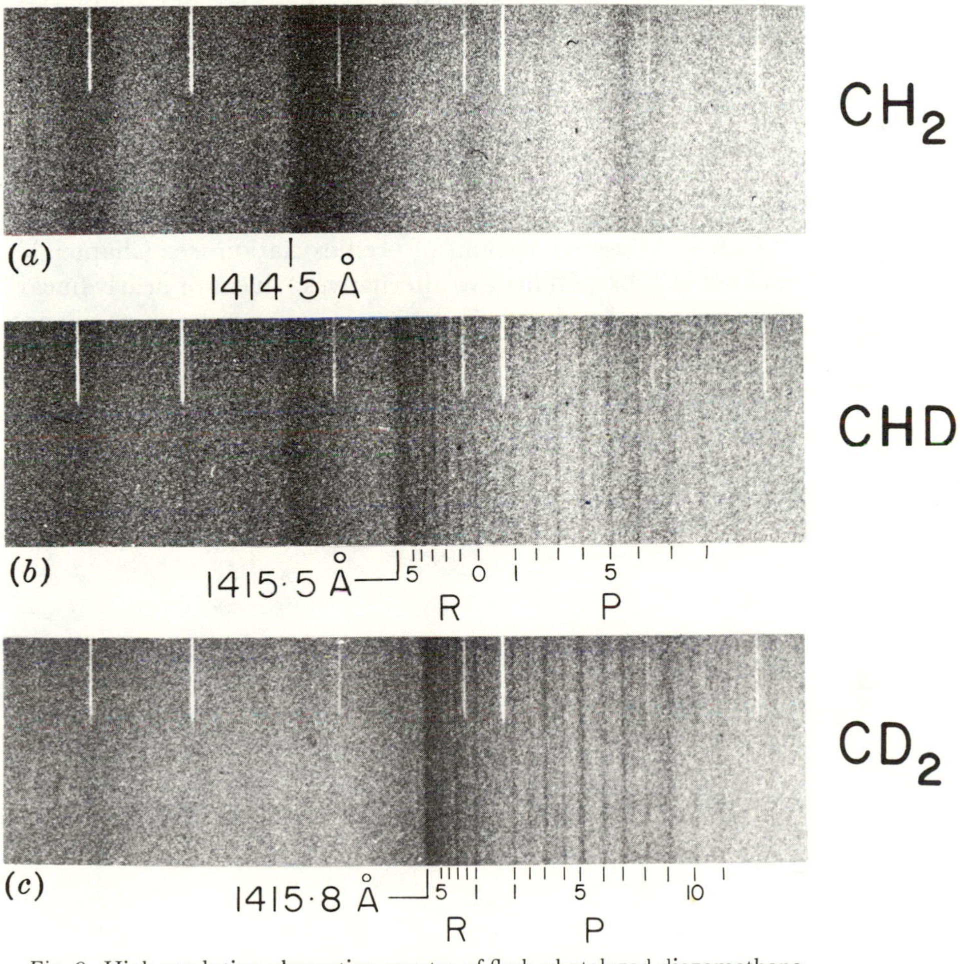

Fig. 9. High-resolution absorption spectra of flash-photolyzed diazomethane and deuterated diazomethanes [from (57)].

Spectrum (c) refers to fully deuterated diazomethane. The intensity alternation evident in the 1414.8 Å band proves that two D atoms are present in symmetrical positions; that is, that this band is due to CD_2. Spectrum (b) was taken with 50 percent deuterated diazomethane. In it the absorption band of CHD is most prominent; it shows no intensity alternation. The lines of CH_2 in spectrum (a) are very diffuse but do show an intensity alternation.

isotopes a clear and simple fine structure with a characteristic intensity alternation for the fully deuterated species. As we shall see in more detail later, this fine structure immediately shows that the molecule responsible contains two and only two H (or D) atoms in symmetrical positions, and indeed that the molecule is linear or nearly linear and can have only one heavier atom in addition to the two H atoms. That this third atom is carbon, while obvious from the nature of the parent compound, was later independently confirmed by experiments with $C^{13}H_2N_2$: the spectrum of $C^{13}H_2$ is slightly shifted compared to that of $C^{12}H_2$. Unlike the spectra of CD_2 and CHD, the spectrum of CH_2 is diffuse on account of predissociation (see Chapter V). [Added in 1988:] Of the two alternatives "linear or nearly linear" just mentioned, subsequent research by several groups of spectroscopists has shown that the second alternative is correct and that the angle H–C–H is approximately 136°. (See the review in McKellar et al. [143] and Bunker et al. [144].)

II. DIATOMIC RADICALS AND IONS

A. ENERGY LEVELS AND EIGENFUNCTIONS

In a first approximation, the energy of a molecule can be represented as the sum of three parts—the electronic, the vibrational, and the rotational energy:

$$E = E_e + E_v + E_r. \tag{1}$$

In Fig. 10 the vibrational and rotational levels in two electronic states are represented schematically. Corresponding to the three

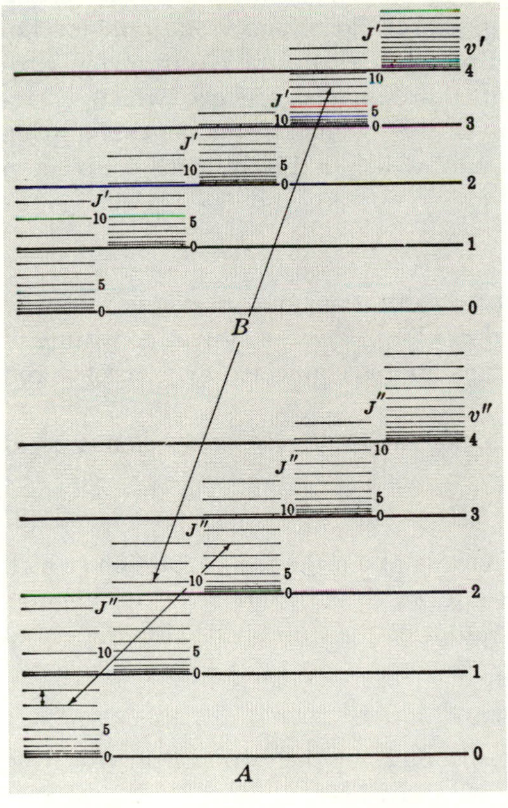

Fig. 10. Vibrational and rotational levels of two electronic states A and B of a molecule (schematic).

The three double arrows indicate examples of transitions in the pure rotation spectrum, the rotation-vibration spectrum, and the electronic spectrum of the molecule.

kinds of energy, we have three kinds of spectra: (a) *rotation spectra*, in which transitions take place from the rotational levels of a given vibrational level in a given electronic state to other rotational levels of the same vibrational and electronic state; (b) *rotation-vibration spectra*, in which transitions take place from the rotational levels of a given vibrational level to the rotational levels of another vibrational level of the same electronic state; and (c) *electronic spectra*, in which transitions from the rotational levels of the various vibrational levels of one electronic state take place to the rotational and vibrational levels of another electronic state.

In addition to these three types of spectra, in recent years several spectra have been observed in the radio-frequency and microwave regions that correspond to transitions between certain fine-structure levels of a given rotational and vibrational level of a given electronic state. Special cases are the electron-spin resonance and nuclear-magnetic-resonance spectra, which correspond to transitions between the Zeeman components of a given level (the components into which a given level splits in a magnetic field).

(1) Rotation

The simplest model for the rotations of a diatomic molecule or radical is the so-called *dumbbell model*—a system consisting of two mass points of masses m_1 and m_2 connected by a massless rod of length r.

In classical mechanics the *energy of rotation* of such a rigid rotator is given by

$$E_r = \tfrac{1}{2}Iw^2, \qquad (2)$$

where I is the moment of inertia about the axis of rotation (which is perpendicular to the line m_1——m_2) and w is the angular velocity. The *moment of inertia* of the dumbbell with respect to the centre of mass is given by

$$I = m_1r_1^2 + m_2r_2^2 = \mu r^2,$$

where r_1 and r_2 are the distances of the two masses from the centre of mass ($r = r_1 + r_2$) and where

$$\mu = \frac{m_1 m_2}{m_1 + m_2} \tag{3}$$

is the *reduced mass*. The *angular momentum* of the system is given by

$$P = Iw.$$

Introducing this into the equation for the energy, we obtain

$$E_r = \frac{P^2}{2I}. \tag{4}$$

In wave mechanics only certain discrete energy levels are possible, which we may obtain by solving the Schrödinger equation of the rigid rotator. The detailed calculation may be found, for example, in Pauling and Wilson (110). The result is

$$E_r = \frac{h^2}{8\pi^2 \mu r^2} J(J+1) = \frac{h^2}{8\pi^2 I} J(J+1), \tag{5}$$

where J, the *rotational quantum number*, can assume the values 0, 1, 2, ... This quantum number is related to the magnitude of the angular-momentum vector $\boldsymbol{P}$ by the relation

$$|\boldsymbol{P}| = \frac{h}{2\pi} \sqrt{J(J+1)}. \tag{6}$$

If this is substituted into the classical expression (4), the energy expression (5) immediately follows. In molecular spectroscopy the energy equation (5) is usually written in the form of *term values*, which are the energy values divided by hc and are measured in cm^{-1}. We have

$$F(J) = \frac{E_r}{hc} = BJ(J+1), \tag{7}$$

where

$$B = \frac{h}{8\pi^2 cI} = \frac{h}{8\pi^2 c\mu r^2} \tag{8}$$

is called the *rotational constant* and is, apart from a constant factor, the reciprocal moment of inertia.

To each of the energy values (eigenvalues) of the wave equation correspond characteristic functions (eigenfunctions) whose squares give the probability distributions. For the rigid rotator these eigenfunctions are the so-called surface harmonics, which are represented in Fig. 11. For each value of $J > 0$ there are several functions corresponding to different values of the magnetic quantum number M, which takes the values

$$M = J, J - 1, \ldots, -J. \tag{9}$$

$Mh/2\pi$ is the component of the angular momentum in the direction of a magnetic (or electric) field applied in the z direction.

If the molecule is not completely rigid, there will be, on account of the centrifugal force, a slight increase in the internuclear distance when the molecule is rotating. As a result the rotational term-value is somewhat modified and takes the form

$$F(J) = BJ(J+1) - DJ^2(J+1)^2 + \cdots, \tag{10}$$

where B is given as before by Eq. (8) and D is a small correction term, always much smaller than B, approximated by

$$D = \frac{4B^3}{\omega^2}, \tag{11}$$

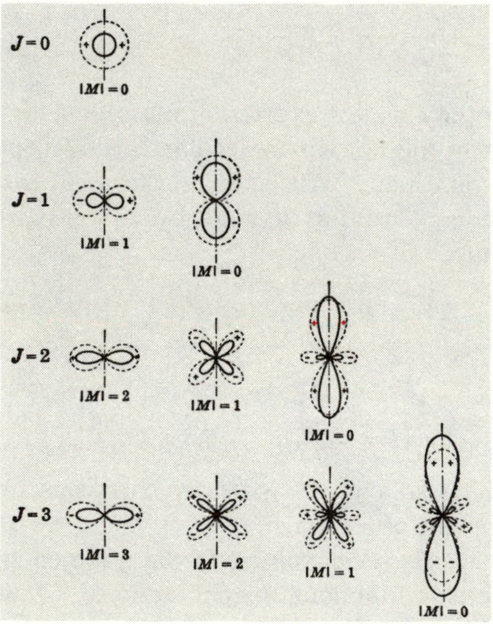

Fig. 11. Eigenfunctions (broken curves) and probability distributions (solid curves) for the rotator in the levels $J = 0, 1, 2, 3$ (from *MM* I, p. 70).

The values of ψ_r and $|\psi_r|^2$ are plotted in polar diagrams.

where ω is the vibrational frequency in cm^{-1} [see subsection (2)]. For example, in the ground state of the C_2 radical we have for the lowest vibrational level ($v = 0$) the values

$$B_0 = 1.811 \text{ cm}^{-1}, \qquad D_0 = 7 \times 10^{-6} \text{ cm}^{-1},$$

while in the ground state of the CH radical

$$B_0 = 14.19 \text{ cm}^{-1}, \qquad D_0 = 14.4 \times 10^{-4} \text{ cm}^{-1}.$$

(2) Vibration

In a first approximation the vibrations of a diatomic molecule can be represented by the model of the *harmonic oscillator*. A harmonic oscillator is a mechanical system consisting of a mass point under the action of a restoring force that is proportional to the displacement x of the mass point from its equilibrium position. The motion of the two nuclei of the molecule can be reduced to the motion of a single particle of mass μ [see Eq. (3)] if the change $r - r_e$ of the internuclear distance from its equilibrium value r_e is equated to the displacement x of the oscillator from its equilibrium position. The potential energy of this harmonic oscillator (molecule) is

$$V = \tfrac{1}{2}kx^2 = \tfrac{1}{2}k(r - r_e)^2. \qquad (12)$$

Here the *force constant* k is related to the vibrational frequency of the oscillator ν_{osc} by the relation

$$k = 4\pi^2\mu\nu_{osc}^2. \qquad (13)$$

If the potential energy (12) is substituted into the wave equation, one finds for the energy levels of the harmonic oscillator

$$E_v = h\nu_{osc}(v + \tfrac{1}{2}),$$

where the vibrational quantum number v can take the values 0, 1, 2, 3, ...

In term values, the vibrational energy formula becomes

$$G(v) = \frac{E_v}{hc} = \omega\left(v + \frac{1}{2}\right), \qquad (14)$$

where

$$\omega = \frac{\nu_{osc}}{c} \qquad (15)$$

is the vibrational frequency measured in wave-number units (cm^{-1}).

The eigenfunctions of the harmonic oscillator are the Hermite orthogonal functions [see Pauling and Wilson (110)], which for the first few v values are represented in Fig. 12. It should be noted that even in the lowest vibrational level, $v = 0$, the vibrational energy is not zero but $\frac{1}{2}\omega$. The corresponding wave function has a bell-like shape; it is a Gaussian curve as given at the bottom of Fig. 12.

In actual fact the molecule is not strictly a harmonic oscillator

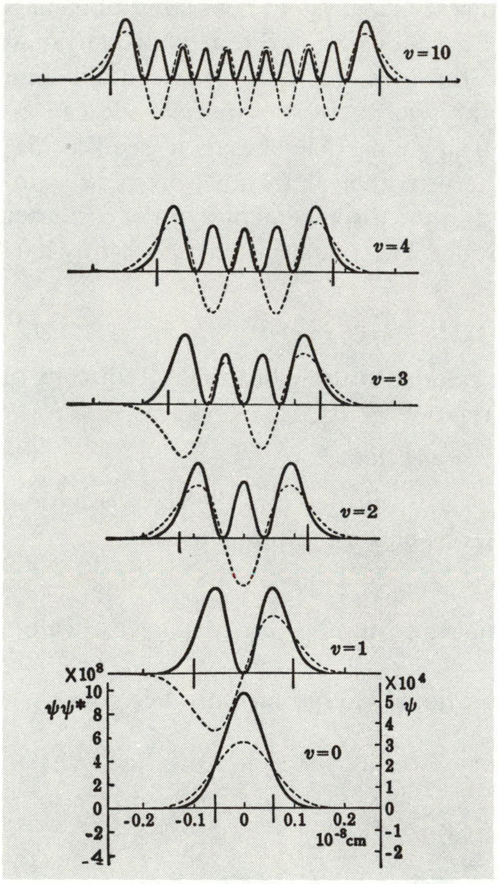

Fig. 12. Eigenfunctions (broken curves) and probability distributions (solid curves) of the harmonic oscillator for $v = 0, 1, 2, 3, 4,$ and 10 (from *MM* I, p. 77).

The abscissae give the displacements from the equilibrium position.

but an *anharmonic oscillator;* that is, the potential function is not a parabola but a curve of the type given in Fig. 13, which near the minimum can be represented by adding to Eq. (12) cubic and higher-power terms. In order to approximate the whole of

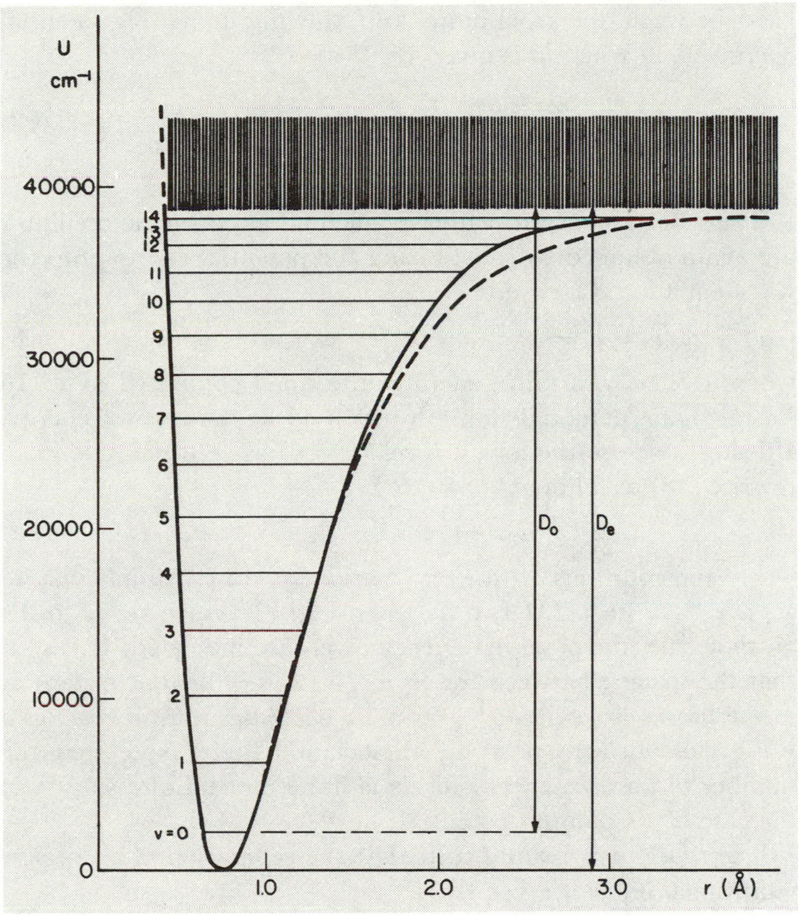

Fig. 13. Potential function and vibrational energy levels of an anharmonic oscillator.

The figure shows the observed vibrational levels in the ground state of the H_2 molecule [Herzberg and Howe (62)] and the potential curve derived from them by Weissman, Vanderslice, and Battino (136). The broken-line curve is the corresponding Morse function, Eq. (16). The hatched area corresponds to the continuous range of energy levels above the asymptote.

the course of the potential function one often uses the *Morse function*, which is given by

$$V = D_e[1 - e^{-\beta(r-r_e)}]^2, \tag{16}$$

where D_e is the dissociation energy—that is, the energy difference between the asymptote and the minimum of potential energy—and where β is given by

$$\beta = \sqrt{\frac{2\pi^2 c\mu}{D_e h}}\,\omega_e = 1.2177_7 \times 10^7 \omega_e \sqrt{\frac{\mu_A}{D_e}} \tag{17}$$

(μ_A is the reduced mass in atomic mass units).

The vibrational term-values of such an anharmonic oscillator are again obtained by substituting the potential energy into the wave equation. One finds

$$G(v) = \omega_e(v + \tfrac{1}{2}) - \omega_e x_e(v + \tfrac{1}{2})^2 + \omega_e y_e(v + \tfrac{1}{2})^3 + \cdots, \tag{18}$$

where $\omega_e x_e$, $\omega_e y_e$ are in general quite small compared to ω_e. In Eq. (18) the potential minimum is used as the zero of energy. If instead we use the lowest level ($v = 0$) as zero point, we can rewrite (18) as follows:

$$G_0(v) = \omega_0 v - \omega_0 x_0 v^2 + \omega_0 y_0 v^3 + \cdots, \tag{18a}$$

where the constants ω_0, $\omega_0 x_0$, . . . are related in a simple way to ω_e, $\omega_e x_e$, . . . (see *MM* I, p. 93).[1] In Fig. 13, which refers to the H_2 molecule, the observed energy levels are indicated. It is seen that the spacing between the levels, instead of being constant as in the harmonic oscillator, gradually decreases until it reaches a value close to zero near the dissociation limit. Note that the number of possible energy levels is finite, not infinite as it is in the case of a Coulomb potential.

From (18) it is readily seen that the separation of successive vibrational levels is given by

$$\begin{aligned}\Delta G(v + \tfrac{1}{2}) &= G(v + 1) - G(v)\\ &= (\omega_e - \omega_e x_e + \omega_e y_e + \cdots) - (2\omega_e x_e - 3\omega_e y_e - \cdots)(v + \tfrac{1}{2})\\ &\qquad + 3\omega_e y_e(v + \tfrac{1}{2})^2 + \cdots;\end{aligned} \tag{19}$$

[1] *MM* I refers to the author's *Molecular Spectra and Molecular Structure*, I, *Diatomic Molecules*, 2nd ed. (Van Nostrand Reinhold Company, a division of Litton Educational Publishing, Inc., New York, 1950); similarly, *MM* II and *MM* III refer to volumes II (1945) and III (1966) of this series.

that is, it decreases slowly with increasing v. The eigenfunctions of the anharmonic oscillator are similar to those of the harmonic oscillator (Fig. 12) but somewhat distorted (see *MM* I, p. 94).

In general, rotation and vibration of the molecule will take place simultaneously, and therefore the effect of the interaction between these two motions must be considered. If the molecule is vibrating, the internuclear distance is changing; therefore the moment of inertia is changing. As a result the rotational constant B_v for a given vibrational level v is different from the rotational constant for the equilibrium position, B_e. To find B_v we must average over the various values of the internuclear distance; that is, we must find

$$B_v = \frac{h}{8\pi^2 c\mu}\left[\frac{1}{r^2}\right]_{\mathrm{av}}, \tag{20}$$

which can be shown to yield

$$B_v = B_e - \alpha_e(v + \tfrac{1}{2}) + \cdots. \tag{21}$$

Here

$$B_e = \frac{h}{8\pi^2 c\mu r_e^2}, \tag{22}$$

while α_e is a constant (small compared to B_e) that depends on the anharmonicity of the vibration as well as on B_e and ω_e.

In a similar way one finds that the rotational constant D that expresses the effect of the centrifugal force is given by

$$D_v = D_e + \beta_e(v + \tfrac{1}{2}) + \cdots, \tag{23}$$

where D_e is the value of this constant for the equilibrium position.

According to the preceding discussion the rotational term-values for a given vibrational level, when interaction of rotation and vibration is taken into account, are no longer given by Eq. (10) but by

$$F_v(J) = B_vJ(J+1) - D_vJ^2(J+1)^2 + \cdots, \tag{24}$$

in which the subscript v refers to the particular vibrational level considered.

(3) Electronic states and electronic eigenfunctions

In order to understand the occurrence of different electronic states with different rotational and vibrational constants we must consider the motion of the electrons around the two nuclei and study their energies. Consider first a single electron in the field of two fixed nuclei. The H_2^+ ion would be an example. The solution of the Schrödinger equation of such a system shows that all positive energy values are possible but only certain negative energy values, just as in the case of the hydrogen atom. Qualitatively we can obtain the discrete (negative) energy levels from those of the so-called *united atom* (He^+ in the case of H_2^+) by applying a strong electric field. In the united atom the electron is characterized by the quantum numbers n and l, the principal quantum number and the azimuthal quantum number. The latter indicates the orbital angular momentum in units $h/2\pi$. As for other angular-momentum vectors in quantum theory one uses $\boldsymbol{l}$ as a symbol for the electronic orbital angular momentum and l for the corresponding quantum number; the magnitude of $\boldsymbol{l}$ is

$$\sqrt{l(l+1)}\,\frac{h}{2\pi} \approx l\left(\frac{h}{2\pi}\right).$$

In an electric field (such as would exist between two nuclei) the orbital angular momentum vector $\boldsymbol{l}$ can only have those orientations with regard to the field direction for which its component in the field direction is $m_l h/2\pi$, where

$$m_l = l,\ l-1,\ l-2,\ \ldots,\ -l. \tag{25}$$

In an electric field the energy is given in a first approximation by

$$E = Cm_l^2;$$

that is, states of the electron differing only in the sign of m_l have the same energy. Therefore, to distinguish states of the electron (orbitals) of given n and l and different $|m_l|$ one uses in place of m_l the quantum number

$$\lambda = |m_l| = l,\ l-1,\ \ldots,\ 0. \tag{26}$$

The orbital wave functions of one-electron states with $\lambda = 0, 1, 2, \ldots$ are called, for short, $\sigma, \pi, \delta, \ldots$ *orbitals* and the electrons in such orbitals $\sigma, \pi, \delta, \ldots$ electrons, in analogy to the corresponding designation in atoms, where the letters $s, p, d, \ldots$ are used to denote electrons or orbitals with $l = 0, 1, 2, \ldots$. Molecular orbitals with $\lambda \neq 0$ have two component functions (with $m_l = +\lambda$ and $m_l = -\lambda$): they are *doubly degenerate;* orbitals with $\lambda = 0$ (σ orbitals) have only one component function: they are *nondegenerate.*

In order to distinguish different $\sigma, \pi, \ldots$ orbitals we can use the n and l values of the corresponding orbitals in the united atom, writing symbolically

$$1s\sigma,\ 2s\sigma,\ 2p\sigma,\ 2p\pi,\ 3s\sigma,\ 3p\sigma,\ 3p\pi,\ 3d\sigma,\ 3d\pi,\ 3d\delta,\ \ldots,$$

where the number gives the n value and $s, p, d, \ldots$ the l value in the united atom. Figure 14 illustrates the shapes of the orbital wave functions corresponding to various n, l, λ values by showing their nodal surfaces.

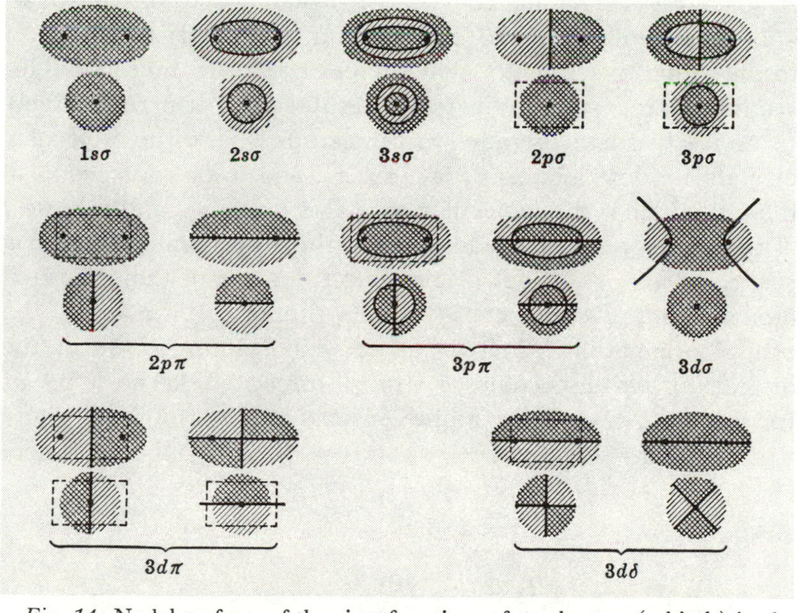

Fig. 14. Nodal surfaces of the eigenfunctions of an electron (orbitals) in the field of two fixed centers (schematic) [after Weizel (137), from *MM* I, p. 326]. In each case two cross sections are shown, one containing the internuclear axis, the other perpendicular to it. The sign of the eigenfunction is indicated by single and cross hatching. For π and δ electrons there are two eigenfunctions corresponding to the double degeneracy.

Instead of using the values of n and l in the united atom, we may also use for a description of a molecular-orbital function the values of these quantum numbers in the separated atoms. It is customary to add these symbols *after* the symbol representing λ and to add as a subscript the nucleus to which they refer. We have therefore symbols such as $\sigma 1s_A$, $\sigma 2p_B$, and so on. It must be noted here that the quantum number λ is the only "good" quantum number which retains its meaning all the way from the united atom to the separated atoms, while the values of n and l for the united atom are meaningful only for small internuclear distances, and those for the separated atoms are meaningful only for large internuclear distances and are in general different from those for the united atom.

If the two nuclei in the molecule have the same charge (that is, for a *homonuclear* molecule), the orbital wave functions can only be symmetric (even) or antisymmetric (odd) with respect to the center of symmetry that then exists in the molecule. This symmetry property is indicated by a subscript g or u, respectively (from the German "gerade" or "ungerade"), and the molecular orbitals are described as σ_g, σ_u, π_g, π_u, and so on. For even l of the united atom the molecular orbital is even (g), and for odd l of the united atom it is odd (u). On the other hand, for each σ (or π, or δ, . . .) orbital originating from one of the separated atoms with a given l there is another similar σ (or π, or δ, . . .) orbital originating from the other (equal) atom. One of the orbitals of a pair becomes σ_g (or π_g, or δ_g, . . .), the other σ_u (or π_u, or δ_u, . . .). For example, instead of the orbitals $\sigma 1s_A$ and $\sigma 1s_B$ when $A \neq B$, we have, when $A = B$, $\sigma_g 1s$ and $\sigma_u 1s$, where

$$\sigma_g 1s = \frac{1}{\sqrt{2}} (\sigma 1s_A + \sigma 1s_B),$$

$$\sigma_u 1s = \frac{1}{\sqrt{2}} (\sigma 1s_A - \sigma 1s_B);$$

that is, both $\sigma_g 1s$ and $\sigma_u 1s$ contain equal contributions from the two identical atoms. (The factor $1/\sqrt{2}$ is introduced in order to normalize $\sigma_g 1s$ and $\sigma_u 1s$, assuming that $\sigma 1s_A$ and $\sigma 1s_B$ are normalized.)

In Fig. 15 are presented schematically the energy levels of a one-electron system in which the designations corresponding to the united atom are given.

In a many-electron system each electron, in a rough approximation, may be considered separately as moving in an approximately axial field of the nuclei and the other electrons. Each electron in this rough approximation can be described by the quantum numbers n_i, l_i, and λ_i, where n_i and l_i refer to either the united atom or the separated atoms. The wave function ψ

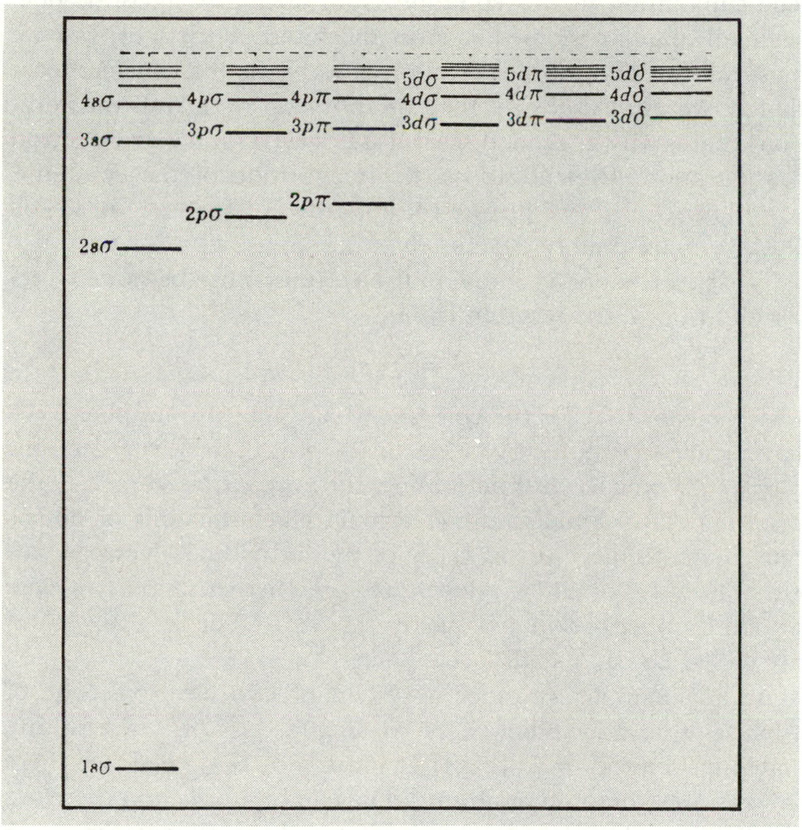

Fig. 15. Energy levels of a single electron in the field of two fixed centers at a small distance from each other (schematic) (from *MM* I, p. 324).

The broken line at the top represents the ionization limit. Further series of levels with higher values of l would join on to the right.

of such a many-electron system in this approximation is simply the product of the individual orbital wave functions $\chi_i(q_i)$:

$$\psi = \chi_1(q_1)\chi_2(q_2)\chi_3(q_3) \ldots \tag{27}$$

For a given electron configuration we obtain in general several (and at least one) electronic states. Each of these states is characterized by a *total orbital angular momentum* $\boldsymbol{\Lambda}$ (in units $h/2\pi$), which is related to the $\boldsymbol{\lambda}_i$ simply by

$$\boldsymbol{\Lambda} = \sum \boldsymbol{\lambda}_i. \tag{28}$$

The summation in this equation is an ordinary algebraic one, since all angular momenta lie in the internuclear axis, but we must take account of the direction of $\boldsymbol{\lambda}_i$ in the internuclear axis; that is, we must use the signed quantity $m_{l_i} = \pm\lambda_i$ in the addition; that is why in general several states arise for a given electron configuration. Depending on the magnitude of the resultant, $\Lambda = 0, 1, 2, \ldots$, we denote the state by Σ, Π, Δ, . . . of which Π, Δ, . . . are doubly degenerate.

The spins $\boldsymbol{s}_i$ of the individual electrons must be added vectorially to give the resultant spin,

$$\boldsymbol{S} = \sum \boldsymbol{s}_i, \tag{29}$$

which means that for the corresponding quantum number S we have, since $s_i = \frac{1}{2}$, for two electrons $S = 1, 0$, for three electrons $S = \frac{3}{2}, \frac{1}{2}, \frac{1}{2}$, and so on. S determines the *multiplicity*, $2S + 1$, of the resulting state. Finally, the electronic eigenfunctions of homonuclear molecules, just as those of the individual electrons, are either even (g) or odd (u) depending on whether there is an even or odd number of odd orbitals (σ_u, π_u, . . .). The resulting states are called Σ_g, Π_g, . . . Σ_u, Π_u . . . .

As an example, consider a system of two electrons, one of which is σ_g and the other π_u. According to Eq. (28), the resulting state must have $\Lambda = 1$; that is, it must be a Π state. Since there are two electrons, the resultant spin S can be 1 or 0, and therefore the Π state will occur both as a triplet and as a singlet. Since one electron is g and the other u, the resulting wave function must have the property u. Thus we have $^3\Pi_u$, $^1\Pi_u$ as the states resulting from the electron configuration $\sigma_g\pi_u$. In a similar way we obtain for $\pi_u\delta_g$ the four states $^1\Pi_u$, $^1\Phi_u$, $^3\Pi_u$, $^3\Phi_u$.

For a configuration of two π electrons we obtain $\Lambda = 2$ or 0, corresponding to Δ and Σ states. Both values of Λ arise in two different ways, which may be represented by $\leftarrow\leftarrow$, $\rightarrow\rightarrow$ and $\leftarrow\rightarrow$, $\rightarrow\leftarrow$. While the first two are degenerate in any approximation if the molecule is not rotating, the last two are degenerate only in a low approximation. In higher approximation they split into two states whose wave functions are the sum and the difference of the functions corresponding to $\leftarrow\rightarrow$ and $\rightarrow\leftarrow$. These functions are symmetric and antisymmetric, respectively, with respect to a reflection at any plane through the internuclear axis. We express this symmetry property by writing Σ^+ and Σ^-, respectively. Therefore, proceeding otherwise as before, we obtain from the configuration $\pi\pi$ the states $^1\Sigma^+$, $^1\Sigma^-$, $^1\Delta$, $^3\Sigma^+$, $^3\Sigma^-$, $^3\Delta$.

If there are equivalent electrons—that is, electrons with the same quantum numbers n, l, and λ—we must take account of the *Pauli principle*, which requires[2] that no two electrons can have the same set of four quantum numbers n, l, m_l, and m_s—or, in other words, that no more than two electrons can be in a given σ orbital, and this only if they have antiparallel spins—while π, δ, . . . orbitals can have a maximum of four electrons, since $m_l = \pm\lambda$ and $m_s = \pm\frac{1}{2}$. Thus the configuration $(1s\sigma_g)^2$ gives only one state, $^1\Sigma_g{}^+$. On the other hand the configuration $(2p\pi_u)^2$ gives rise to three states, which can be shown to be $^3\Sigma_g{}^-$, $^1\Delta_g$, $^1\Sigma_g{}^+$—just half the number of states given above for the configuration $\pi\pi$. For a heteronuclear molecule the u and g would have to be dropped.

If we now want to predict what electronic states we might expect for a given diatomic molecule, we need only study the various ways in which we can distribute the electrons over the possible orbitals. We obtain the ground state of a molecule by putting all electrons into the lowest orbitals to the extent allowed by the Pauli principle. For example, in the H_2 molecule the ground state is obtained by putting both electrons in the $1s\sigma_g$ orbital, resulting in a $^1\Sigma_g{}^+$ state. The ground state of BeH with five electrons is obtained by putting two electrons each in the $1s\sigma$ and $2s\sigma$ orbitals and the fifth electron into the next lowest orbital $2p\sigma$, resulting in a $^2\Sigma^+$ state. In a similar way the ground

[2] See, for example, *AA*, p. 120.

states of other diatomic hydrides are obtained, as given in Table 1. The excited states are obtained by taking an electron to higher unoccupied or only partially filled orbitals. In this

Table 1. Electron Configurations and Term Types of the Lowest States of Diatomic Hydrides

Molecule	*Lowest Electron Configuration*	*First Excited Electron Configuration*
LiH, BeH^+	$K(2s\sigma)^2\ {}^1\Sigma^+$	$2s\sigma 2p\sigma\ {}^1\Sigma^+\ [{}^3\Sigma^+]$
NaH, MgH^+	$KL(3s\sigma)^2\ {}^1\Sigma^+$	$3s\sigma 3p\sigma\ {}^1\Sigma^+\ [{}^3\Sigma^+]$
KH, CaH^+	$KLM_{sp}(4s\sigma)^2\ {}^1\Sigma^+$	$4s\sigma 3d\sigma\ {}^1\Sigma^+\ [{}^3\Sigma^+]$
CuH, ZnH^+	$KLM(4s\sigma)^2\ {}^1\Sigma^+$	$4s\sigma 4p\sigma\ {}^1\Sigma^+\ [{}^3\Sigma^+]$
RbH	$KLMN_{sp}(5s\sigma)^2\ {}^1\Sigma^+$	$5s\sigma 4d\sigma\ {}^1\Sigma^+\ [{}^3\Sigma^+]$
BeH, BH^+	$K(2s\sigma)^2\ 2p\sigma\ {}^2\Sigma^+$	$(2s\sigma)^2\ 2p\pi\ {}^2\Pi_r$
MgH, AlH^+	$KL(3s\sigma)^2\ 3p\sigma\ {}^2\Sigma^+$	$(3s\sigma)^2\ 3p\pi\ {}^2\Pi_r$
CaH	$KLM_{sp}(4s\sigma)^2\ 3d\sigma\ {}^2\Sigma^+$	$(4s\sigma)^2\ 4p\pi\ {}^2\Pi_r$
ZnH	$KLM(4s\sigma)^2\ 4p\sigma\ {}^2\Sigma^+$	$(4s\sigma)^2\ 4p\pi\ {}^2\Pi_r$
SrH	$KLMN_{sp}(5s\sigma)^2\ 4d\sigma\ {}^2\Sigma^+$	$(5s\sigma)^2\ 4d\pi\ {}^2\Pi_r$
CdH	$KLMN_{spd}(5s\sigma)^2\ 5p\sigma\ {}^2\Sigma^+$	$(5s\sigma)^2\ 5p\pi\ {}^2\Pi_r$
BH, CH^+	$K(2s\sigma)^2(2p\sigma)^2\ {}^1\Sigma^+$	$(2s\sigma)^2\ 2p\sigma 2p\pi\ {}^1\Pi,\ {}^3\Pi$
AlH	$KL(3s\sigma)^2(3p\sigma)^2\ {}^1\Sigma^+$	$(3s\sigma)^2\ 3p\sigma 3p\pi\ {}^1\Pi,\ {}^3\Pi$
InH	$KLM_{spd}(5s\sigma)^2(5p\sigma)^2\ {}^1\Sigma^+$	$(5s\sigma)^2\ 5p\sigma 5p\pi\ {}^1\Pi,\ {}^3\Pi$
CH	$K(2s\sigma)^2(2p\sigma)^2\ 2p\pi\ {}^2\Pi_r$	$(2s\sigma)^2\ 2p\sigma(2p\pi)^2\ [{}^4\Sigma^-],\ {}^2\Delta,\ {}^2\Sigma^+,\ {}^2\Sigma^-$
SiH	$KL(3s\sigma)^2(3p\sigma)^2\ 3p\pi\ {}^2\Pi_r$	$(3s\sigma)^2\ 3p\sigma(3p\pi)^2\ [{}^4\Sigma^-],\ {}^2\Delta,\ {}^2\Sigma^+,\ [{}^2\Sigma^-]$
SnH	$KLMN_{spd}(5s\sigma)^2(5p\sigma)^2\ 5p\pi\ {}^2\Pi_r$	$(5s\sigma)^2\ 5p\sigma(5p\pi)^2\ {}^4\Sigma^-,\ {}^2\Delta,\ [{}^2\Sigma^+],\ [{}^2\Sigma^-]$
NH, OH^+	$K(2s\sigma)^2(2p\sigma)^2(2p\pi)^2\ {}^3\Sigma^-,\ {}^1\Delta,\ {}^1\Sigma^+$	$(2s\sigma)^2\ 2p\sigma(2p\pi)^3\ {}^3\Pi,\ {}^1\Pi$
PH	$KL(3s\sigma)^2(3p\sigma)^2(3p\pi)^2\ {}^3\Sigma^-,\ {}^1\Delta,\ [{}^1\Sigma^+]$	$(3s\sigma)^2\ 3p\sigma(3p\pi)^3\ {}^3\Pi,\ [{}^1\Pi]$
OH	$K(2s\sigma)^2(2p\sigma)^2(2p\pi)^3\ {}^2\Pi_i$	$(2s\sigma)^2\ 2p\sigma(2p\pi)^4\ {}^2\Sigma^+$
HS, HCl^+	$KL(3s\sigma)^2(3p\sigma)^2(3p\pi)^3\ {}^2\Pi_i$	$(3s\sigma)^2\ 3p\sigma(3p\pi)^4\ {}^2\Sigma^+$
HBr^+	$KLM(4s\sigma)^2(4p\sigma)^2(4p\pi)^3\ {}^2\Pi_i$	$(4s\sigma)^2\ 4p\sigma(4p\pi)^4\ {}^2\Sigma^+$
HF	$K(2s\sigma)^2(2p\sigma)^2(2p\pi)^4\ {}^1\Sigma^+$	$(2s\sigma)^2(2p\sigma)^2(2p\pi)^3\ 3s\sigma\ [{}^3\Pi],\ [{}^1\Pi]$
HCl	$KL(3s\sigma)^2(3p\sigma)^2(3p\pi)^4\ {}^1\Sigma^+$	$(3s\sigma)^2(3p\sigma)^2(3p\pi)^3\ 4s\sigma\ [{}^3\Pi],\ {}^1\Pi$
HBr	$KLM(4s\sigma)^2(4p\sigma)^2(4p\pi)^4\ {}^1\Sigma^+$	$(4s\sigma)^2(4p\sigma)^2(4p\pi)^3\ 5s\sigma\ {}^3\Pi,\ {}^1\Pi$
HI	$KLMN_{spd}(5s\sigma)^2(5p\sigma)^2(5p\pi)^4\ {}^1\Sigma^+$	$(5s\sigma)^2(5p\sigma)^2(5p\pi)^3\ 6s\sigma\ {}^3\Pi,\ {}^1\Pi$

For the excited electron configurations the closed atomic shells have not been repeated. The states in brackets have not been observed. M_{sp} means that in the M shell only the subgroups $3s$ and $3p$ are closed, and similarly in other cases. Π_r and Π_i refer to "regular" and "inverted" Π states; see p. 47.

way we obtain some of the excited states of the radicals given in the second part of Table 1. For example, for BeH the first excited state is obtained by moving the electron from the $2p\sigma$ to the $2p\pi$ orbital, resulting in a $^2\Pi$ state. Of special interest are those states that arise when the most loosely bound electron is taken into orbitals with higher principal quantum numbers (*Rydberg states*). Thus, for example, for CH the states given in Table 2 are obtained.

Table 2. Rydberg States of CH

Electron Configuration	*States*
$1s\sigma^2 2s\sigma^2 2p\sigma^2 2p\pi$	$X\ ^2\Pi$
$1s\sigma^2 2s\sigma^2 2p\sigma^2 3s\sigma$	$^2\Sigma^+$
$1s\sigma^2 2s\sigma^2 2p\sigma 2p\pi 3s\sigma$	$E\ ^2\Pi,\ ^2\Pi,\ ^4\Pi$
$1s\sigma^2 2s\sigma^2 2p\sigma^2 3p\sigma$	$F\ ^2\Sigma$
..........$3p\pi$	$^2\Pi$
$1s\sigma^2 2s\sigma^2 2p\sigma 2p\pi 3p\sigma$	$^2\Pi,\ ^2\Pi,\ ^4\Pi$
.............$3p\pi$	$^2\Sigma^+,\ ^2\Sigma^-,\ ^2\Delta, \ldots$
$1s\sigma^2 2s\sigma^2 2p\sigma^2 3d\sigma$	G { $^2\Sigma^+$
..........$3d\pi$	G { $^2\Pi$
..........$3d\delta$	G { $^2\Delta$
$1s\sigma^2 2s\sigma^2 2p\sigma^2 ns\sigma$	$^2\Sigma^+$
..........$np\sigma$	$^2\Sigma^+$
..........$np\pi$	$^2\Pi$
..........$nd\sigma$	$^2\Sigma^+$
..........$nd\pi$	$^2\Pi$
..........$nd\delta$	$^2\Delta$
$1s\sigma^2 2s\sigma^2 2p\sigma 2p\pi ns\sigma$	$^2\Pi,\ ^2\Pi,\ ^4\Pi$

The states designated by italic letters added to the Greek symbol have been observed [Herzberg and Johns (65)]. In addition some of the states with $n = 4$, 5, 6 have been found.

While the order of the orbitals is fairly obvious in the case of diatomic hydrides (see Figs. 15 and 16), it is not quite so obvious for nonhydrides. In order to find this order, at least in a very rough way, we must establish the correlation between the orbitals at large internuclear distances, where it is determined by the separated atoms, and those at small internuclear distances, where

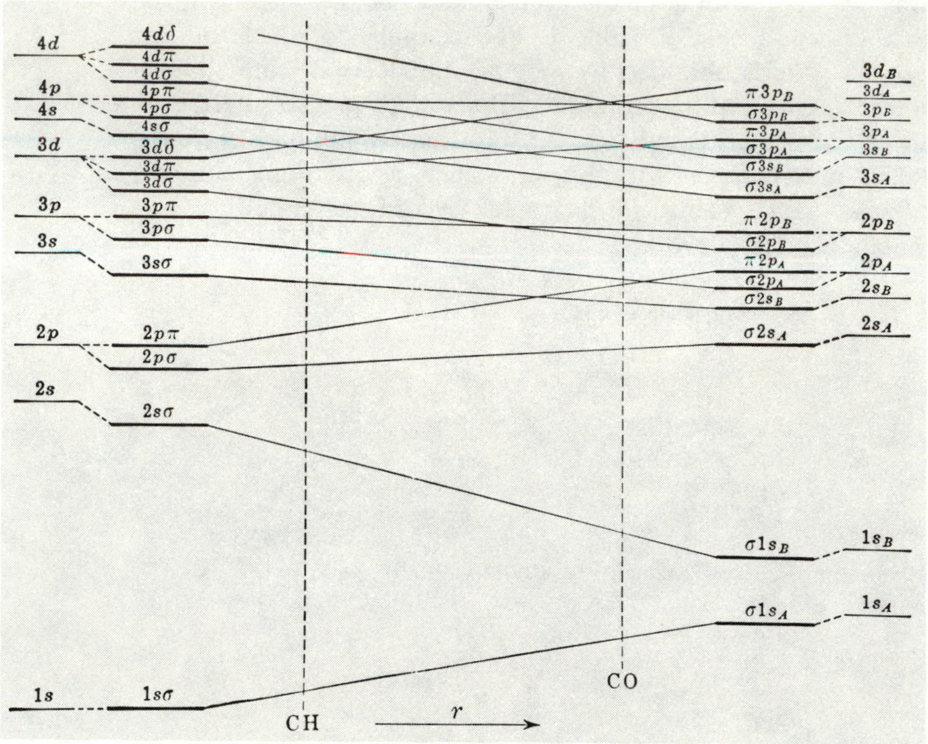

Fig. 16. Correlation of molecular orbitals in a two-center system with unequal nuclear charges (from *MM* I, p. 328).

To the extreme left and the extreme right are given the orbitals in the united and separated atoms, respectively, and, beside them, those in the molecule for very small and very large internuclear distances. The region in between corresponds to intermediate internuclear distances. The vertical broken lines give the approximate positions in the diagram that correspond to the molecules indicated. It should be noticed that the scale of *r* in this and the following figure is by no means linear but becomes rapidly smaller on the right-hand side.

it is determined by the united atom. This is done in Figs. 16 and 17 for the case of unequal and equal nuclear charges, respectively. The correlation must be made in such a way that orbitals of the same type do not intersect. Thus in Fig. 16 the lowest σ orbital on the right goes to the lowest σ orbital on the left. The

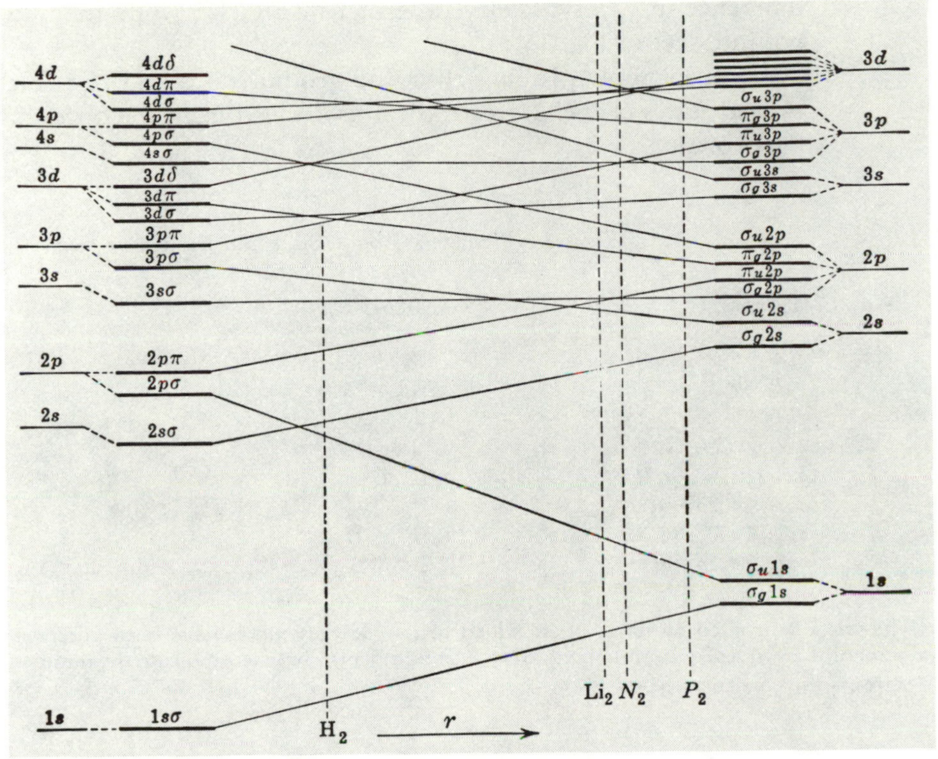

Fig. 17. Correlation of molecular orbitals in a two-center system with equal nuclear charges (from *MM* I, p. 329).
See explanatory note with Fig. 16.

second lowest on the right goes to the second lowest on the left, and so on; and similarly the lowest π orbital on the right goes to the lowest π orbital on the left, and so on. In Fig. 17 we must also consider the symmetry property g and u, which is well defined from large to small internuclear distances, and therefore we must correlate the lowest π_g orbital at the right with the lowest π_g orbital at the left and the lowest π_u orbital on the right with the lowest π_u orbital on the left (that is, $2p\pi$), and so on. We see that the energy curves of certain orbitals, because of the difference in g, u symmetry, intersect in Fig. 17, while they do not intersect in Fig. 16.

As an example, Table 3 lists the ground state and the low excited states of the C_2 free radical as derived from the order of the orbitals in Fig. 17.

Table 3. Predicted Ground State and Excited States of C_2

Electron Configurations	*Resulting States*
$KK(\sigma_g 2s)^2(\sigma_u 2s)^2(\pi_u 2p)^4$	$X\,^1\Sigma_g^+$
.........$(\sigma_u 2s)^2(\pi_u 2p)^3(\sigma_g 2p)$	$a\,^3\Pi_u$, $A^1\Pi_u$
.........$(\sigma_u 2s)^2(\pi_u 2p)^2(\sigma_g 2p)^2$	$b\,^3\Sigma_g^-$, $^1\Delta_g$, $^1\Sigma_g^+$
.........$(\sigma_u 2s)(\pi_u 2p)^4(\sigma_g 2p)$	$c\,^3\Sigma_u^+$, $D^1\Sigma_u^+$
.........$(\sigma_u 2s)(\pi_u 2p)^3(\sigma_g 2p)^2$	$d\,^3\Pi_g$, $C^1\Pi_g$
.........$(\sigma_u 2s)^2(\pi_u 2p)^2(\sigma_g 2p)(\pi_g 2p)$	$e\,^3\Pi_g$, $^3\Pi_g(3)$, $^1\Pi_g(3)$, $^5\Pi_g$, $^3\Phi_g$, $^1\Phi_g$
.........$(\sigma_u 2s)^2(\pi_u 2p)^3(\sigma_g 3s)$	$^3\Pi_u$, $F^1\Pi_u$
.........$(\sigma_u 2s)^2(\pi_u 2p)^2(\sigma_g 2p)(\sigma_g 3s)$	$f\,^3\Sigma_g^-$, $g^3\Delta_g$, $^3\Sigma_g^-$ $^3\Sigma_g^+$, $^5\Sigma_g^-$, $^1\Sigma_g^+$, $^1\Sigma_g^-$, $^1\Delta_g$

The states designated by italic letters added to the Greek symbols have been observed [Herzberg, Lagerqvist, and Malmberg (67)]. The numbers in brackets indicate the number of states of the particular type.

In going from the separated atoms to the molecule, certain general *correlation rules*, first developed by Wigner and Witmer, must hold for the types of the electronic states. If M_{L_1} and M_{L_2} are the magnetic quantum numbers corresponding to the orbital angular momenta $\boldsymbol{L}_1$ and $\boldsymbol{L}_2$ of the separated atoms, then it is immediately clear by the definition of Λ that we must have

$$\Lambda = |M_{L_1} + M_{L_2}|. \tag{30}$$

In addition we must have for the total spin in the molecule

$$\mathbf{S} = \mathbf{S}_1 + \mathbf{S}_2, \tag{31}$$

where $\mathbf{S}_1$ and $\mathbf{S}_2$ are the spins of the separated atoms, or for the corresponding quantum numbers

$$S = S_1 + S_2,\ S_1 + S_2 - 1,\ \ldots,\ |S_1 - S_2|. \tag{32}$$

On this basis it is easy to see, for example, that if a hydrogen atom and a carbon atom both in their ground states[3] ($^2S_g + {}^3P_g$) are brought together, the following molecular states arise: $^2\Pi$, $^2\Sigma^-$, $^4\Pi$, $^4\Sigma^-$. Here the symmetry of the Σ states follows on the basis of a separate rule, which we shall not explain here (refer to *MM* I, p. 318). Table 4 lists several other combinations of

Table 4. Examples of Molecular Electronic States Resulting from Given States of the Separated (Unlike) Atoms

States of Separated Atoms	*Molecular States*
$^2S_g + {}^4S_u$	$^3\Sigma^-$, $^5\Sigma^-$
$^2S_g + {}^2P_u$	$^1\Pi$, $^1\Sigma^+$, $^3\Pi$, $^3\Sigma^+$
$^1S_g + {}^2D_u$	$^2\Sigma^-$, $^2\Pi$, $^2\Delta$
$^2P_g + {}^3P_g$	$^2\Sigma^+(2)$, $^2\Sigma^-$, $^2\Pi(2)$, $^2\Delta$
	$^4\Sigma^+(2)$, $^4\Sigma^-$, $^4\Pi(2)$, $^4\Delta$
$^3P_g + {}^3P_g$	$^1\Sigma^+(2)$, $^1\Sigma^-$, $^1\Pi(2)$, $^1\Delta$
	$^3\Sigma^+(2)$, $^3\Sigma^-$, $^3\Pi(2)$, $^3\Delta$
	$^5\Sigma^+(2)$, $^5\Sigma^-$, $^5\Pi(2)$, $^5\Delta$

For like atoms in unlike states each of these states occurs both as g and as u. For like atoms in like states only one or the other occurs (see *MM* I, p. 321). The numbers in brackets indicate the number of states of the particular type. The same states result when the parities of both atoms are reversed. Change of parity of only one atom changes Σ^- into Σ^+ and Σ^+ into Σ^-.

atoms and the resulting molecular states. It is seen that unless both atoms are in S states, the number of resulting molecular states may be quite large.

Let us now consider the relation between the electronic energy of the molecule determined by all the electrons and the *potential energy* in which the nuclei move. The Schrödinger equation of the molecule is

$$H\psi = E\psi, \tag{33}$$

[3] The g and u added to the atomic term symbols are used here in the same sense as for homonuclear molecules: they indicate the behaviour of the corresponding eigenfunction with respect to inversion. The eigenfunction is "even" (g) when $\sum l_i$ summed over all electrons is even; the eigenfunction is "odd" (u) when $\sum l_i$ is odd. Atomic spectroscopists do not use the g, u notation but indicate "odd" states by a superscript o; they do not specifically indicate the "even" character.

where

$$H = \frac{1}{2m} \sum_i p_i^2 + \frac{1}{2} \sum_k \frac{p_k^2}{M_k} + V. \tag{34}$$

Here the subscripts i refer to the electrons and the subscripts k refer to the nuclei; m is the mass of the electron and M_k the mass of the nucleus k; p is the linear momentum, which in quantum mechanics is replaced by an operator $(h/2\pi i)(\partial/\partial q)$. Substituting these operators, we obtain from (33) and (34)

$$\frac{1}{m} \sum_i \left(\frac{\partial^2 \psi}{\partial x_i^2} + \frac{\partial^2 \psi}{\partial y_i^2} + \frac{\partial^2 \psi}{\partial z_i^2} \right) + \sum_k \frac{1}{M_k} \left(\frac{\partial^2 \psi}{\partial x_k^2} + \frac{\partial^2 \psi}{\partial y_k^2} + \frac{\partial^2 \psi}{\partial z_k^2} \right) + \frac{8\pi^2}{h^2} (E - V)\psi = 0. \tag{35}$$

The potential energy V may be resolved into a sum of an electronic and a nuclear part:

$$V = V_e + V_n. \tag{36}$$

In a first approximation the wave function ψ may be written as a product of a function of the electronic coordinates and a function of the nuclear coordinates—that is,

$$\psi = \psi_e(\ldots, x_i, y_i, z_i, \ldots)\psi_{vr}(\ldots, x_k, y_k, z_k, \ldots). \tag{37}$$

Here the function ψ_e is the solution of the electronic wave equation for a system with two fixed centers,

$$\sum_i \left(\frac{\partial^2 \psi_e}{\partial x_i^2} + \frac{\partial^2 \psi_e}{\partial y_i^2} + \frac{\partial^2 \psi_e}{\partial z_i^2} \right) + \frac{8\pi^2 m}{h^2} (E^{el} - V_e)\psi_e = 0, \tag{38}$$

where E^{el} is the energy of the electrons in the field of the two nuclei. The function ψ_{vr} $(= \psi_v\psi_r$; see below) is the solution of the equation

$$\sum_k \frac{1}{M_k} \left(\frac{\partial^2 \psi_{vr}}{\partial x_k^2} + \frac{\partial^2 \psi_{vr}}{\partial y_k^2} + \frac{\partial^2 \psi_{vr}}{\partial z_k^2} \right) + \frac{8\pi^2}{h^2} (E - E^{el} - V_n)\, \psi_{vr} = 0. \tag{39}$$

If Eq. (37) is substituted into Eq. (35), and if account is taken of Eqs. (36), (38), and (39), one finds that Eq. (35) is fulfilled only if

$$\sum_k \frac{2}{M_k} \left[\frac{\partial \psi_e}{\partial x_k} \frac{\partial \psi_{vr}}{\partial x_k} + \frac{\partial \psi_e}{\partial y_k} \frac{\partial \psi_{vr}}{\partial y_k} + \frac{\partial \psi_e}{\partial z_k} \frac{\partial \psi_{vr}}{\partial z_k} + \frac{1}{2} \psi_{vr} \left(\frac{\partial^2 \psi_e}{\partial x_k^2} + \frac{\partial^2 \psi_e}{\partial y_k^2} + \frac{\partial^2 \psi_e}{\partial z_k^2} \right) \right] \tag{40}$$

can be neglected—that is, if the variation of ψ_e with the nuclear coordinates is slight. This is essentially the so-called *Born-Oppenheimer approximation.* In this approximation, according to Eq. (39), the potential energy under which the nuclei move is obtained simply by adding to the purely electronic energy E^{el} the potential energy V_n of the nuclei, which is given by $Z_1 Z_2 e^2/r$.

Calculations of E^{el}—that is, of the stability of electronic states on the basis of the wave equation—are, of course, not simple.[4] As a very rough guide, we can use correlation diagrams such as those in Figs. 16 and 17, since in these diagrams orbitals whose energies go down in going from right to left will contribute to the binding if they are filled with one or more electrons, while orbitals that go up will contribute to repulsion when filled with one or more electrons. The former electrons (orbitals) are called *bonding electrons* (orbitals), the latter *antibonding electrons* (orbitals). There are also cases where the energy remains more or less the same, in which case the electrons (orbitals) are called *nonbonding electrons* (orbitals).

(4) Coupling of rotation and electronic motion

The rotational levels of a diatomic molecule are characterized by certain *overall symmetry properties*. One of the most important of these is the *parity:* a rotational level is called "positive" (+) or "negative" (−) depending on whether the overall eigenfunction remains unchanged or changes sign for a reflection of all particles at the origin—or, for short, an inversion (that is, for a transition from a right-handed to a left-handed coordinate system). Since the overall eigenfunction can be written as a product

$$\psi = \psi_e \psi_v \psi_r, \tag{41}$$

the overall symmetry will depend only on the rotational eigenfunction ψ_r if the electronic and vibrational functions ψ_e and ψ_v are symmetric with regard to this inversion. The rotational function remains unchanged or changes sign for a reflection at the origin, depending on whether the rotational quantum number J is even or odd (see Fig. 11). Thus for a Σ^+ state the rotational levels for $J = 0, 1, 2, 3, \ldots$ are $+, -, +, -, \ldots$, respectively. On the other hand, for a Σ^- state for which the electronic func-

[4] See *MM* I and *MM* III and the more detailed presentations by Pauling (109), Coulson (23), Parr (108), Murrell, Kettle, and Tedder (101), Streitwieser (125), Daudel (27), Hartmann (48).

tion ψ_e changes sign upon reflection at the origin the overall symmetries are reversed, and we have for $J = 0, 1, 2, 3, \ldots$ the parities $-, +, -, +, \ldots$, respectively. Electronic states with $\Lambda \neq 0$—that is, Π, Δ, . . . states—are doubly degenerate because of the two possible orientations of $\boldsymbol{\Lambda}$ along the axis. Therefore, for every value of J there is both a positive and a negative rotational level. In multiplet states the levels that differ only by the orientation of the spin have the same parity. Figure 18(a) illustrates diagrammatically the parities of the rotational levels for the more important types of electronic states.

For molecules with identical nuclei (homonuclear molecules) we have an additional overall symmetry property, since an exchange of the two nuclei leaves the system unchanged: the overall wave function can only be *symmetric* (s) or *antisymmetric* (a), that is, can only remain unchanged or change sign for such exchange of the nuclei. In an even electronic state (Σ_g, Π_g, and so on; see p. 30) the positive rotational levels are s, the negative rotational levels are a, while for odd electronic states (Σ_u, Π_u, and so on) the positive rotational levels are a and the negative levels are s. If the identical nuclei have zero spin, only the s levels occur. This is the case, for example, for the C_2 free radical. For nonzero nuclear spin I both s and a rotational levels occur, but with different statistical weights their ratio being $(I + 1)/I$ or $I/(I + 1)$ depending on whether I is integral or half-integral. In Fig. 18(b) the symmetry properties $+$, $-$ as well as s, a are shown for a few types of electronic states.

The presence of the electrons in the molecule implies that the moment of inertia I_A about the internuclear axis is not zero, although clearly it is extremely small. Strictly speaking, therefore, the system we are considering is a prolate symmetric top (see p. 151), with one very small and two large and equal principal moments of inertia. The energy formula for such a system is

$$F_v(J) = B_vJ(J + 1) + (A - B_v)\Lambda^2, \tag{42}$$

where

$$A = \frac{h}{8\pi^2 cI_A} \tag{43}$$

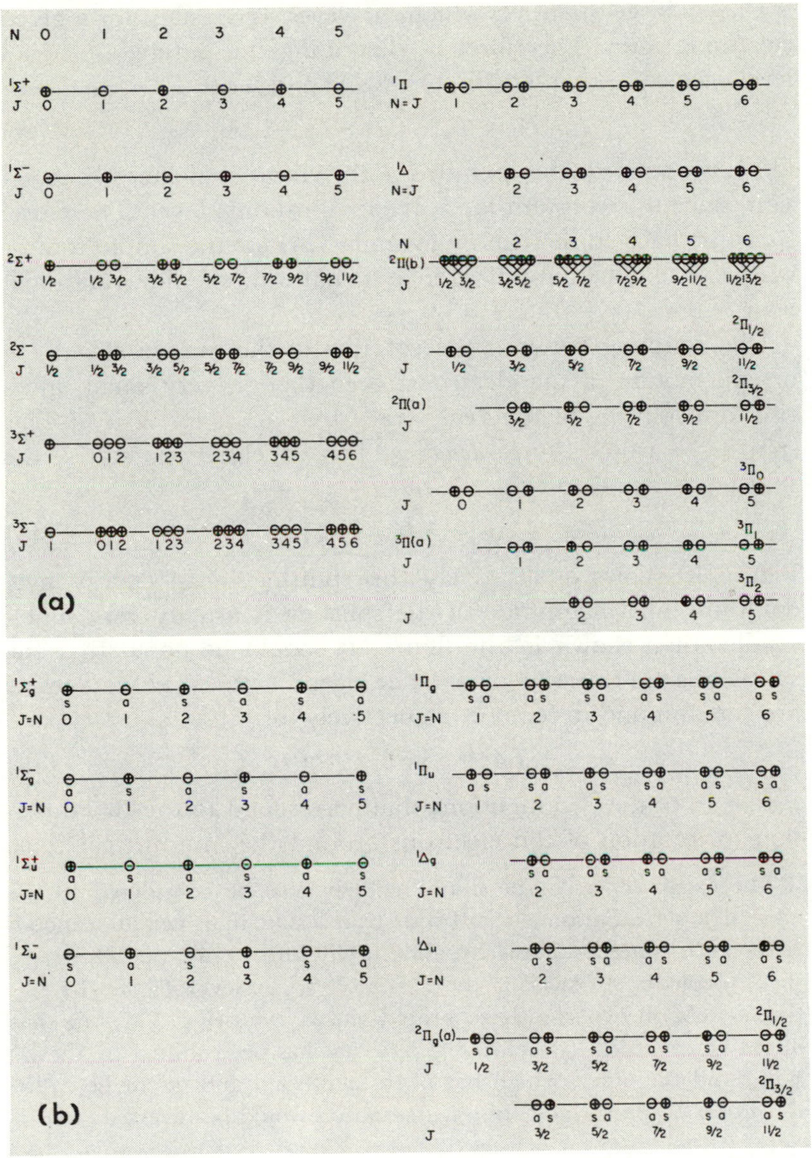

Fig. 18. Symmetry properties of rotational levels in various types of electronic states (a) for heteronuclear molecules (b) for homonuclear molecules. ⊕ stands for positive, ⊖ for negative rotational levels; *s* means symmetric and *a* means antisymmetric.

is a fairly large quantity, which, however, is constant for a given electronic state. Therefore, in describing the purely rotational levels, we can leave out the term $A\Lambda^2$ and write

$$F_v(J) = B_v[J(J+1) - \Lambda^2]. \tag{44}$$

Often the term $B_v\Lambda^2$ is included in the vibrational energy expression, since it is constant for a given vibrational level. Therefore, we come back to the energy formula (24) for the simple rotator with $\Lambda = 0$. The only difference is that now the first rotational level is not $J = 0$, but $J = \Lambda$.

The coupling between the rotation of the molecule and the orbital motion of the electrons, even though very small, gives rise to a splitting of the degeneracy that arises for $\Lambda \neq 0$. This splitting is called Λ-*type doubling*. For $^1\Pi$ electronic states, the splitting is found to be given by

$$\Delta\nu = qJ(J+1). \tag{45}$$

Figure 19 shows qualitatively this splitting in an energy-level diagram. The magnitude of the splitting is usually very much smaller than shown in the figure. It should be noted that the split levels are either + or −. The eigenfunctions for these levels are the sum and difference, respectively, of

$$\chi e^{+i\Lambda\phi} \quad \text{and} \quad \bar{\chi} e^{-i\Lambda\phi} \tag{46}$$

—that is, of the two functions that correspond to the two directions of rotation of the electrons around the internuclear axis.

The existence of Λ-type doubling may also be considered as the effect of a perturbation of the Π state by a Σ state in its neighbourhood. If the perturbing Σ state is Σ^+, then it will affect only the −, +, −, +, . . . component levels of the $J = 1, 2, 3, \ldots$ levels of the Π state; thus we obtain two slightly separated sets of levels (Fig. 19): one that might be called Π^+ (which is the one that has been shifted by the Σ^+ state) and the other, which has been unaffected and might be called Π^-. For a Σ^- perturbing state the situation would be reversed.

The interaction of the electron spin with the orbital angular momentum of the electrons (spin-orbit coupling) causes a further splitting. Without rotation the states $^2\Sigma$, $^3\Sigma$, . . . are unsplit, just as are ^2S, ^3S, . . . states of atoms. On the other hand $^2\Pi$, $^3\Pi$, $^4\Pi$

states, and similarly $^2\Delta$, $^3\Delta$, . . . states, are split into $2S + 1$ components. These components may be distinguished by a quantum number Σ, which represents the component of the spin in the direction of the internuclear axis. It takes the values

$$\Sigma = S, S - 1, \ldots, -S. \tag{47}$$

The resultant of the electronic orbital angular momentum and the spin in the direction of the internuclear axis is called $\boldsymbol{\Omega}$.

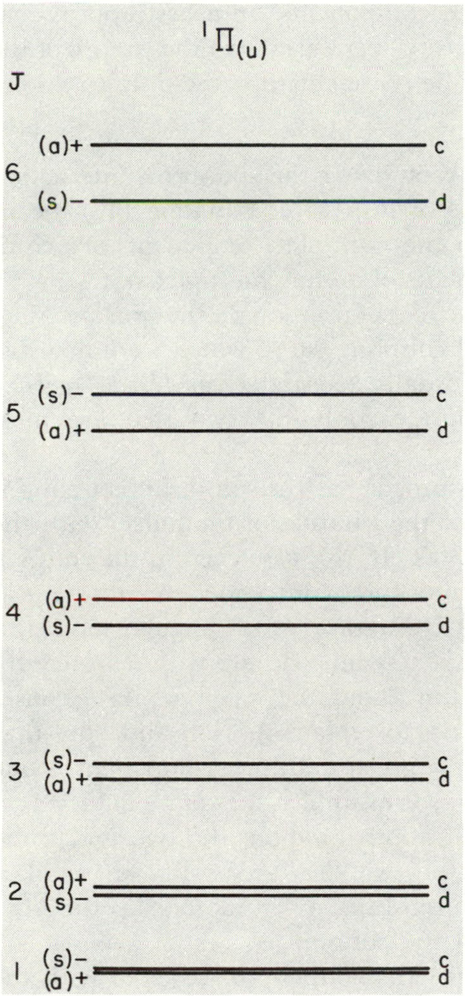

Fig. 19. Λ-type doubling for a $^1\Pi$ state (schematic).

The magnitude of the doubling compared to the rotational spacing is usually very much smaller than shown here.

Its magnitude is $\Omega h/2\pi$, where the quantum number Ω is given by

$$\Omega = \Lambda + \Sigma. \tag{48}$$

The value of Ω is often indicated as a subscript following the term symbol. Thus we have $^2\Pi_{1/2}$, $^2\Pi_{3/2}$ states, $^3\Pi_0$, $^3\Pi_1$, $^3\Pi_2$ states, and $^4\Pi_{5/2}$, $^4\Pi_{3/2}$, $^4\Pi_{1/2}$, $^4\Pi_{-1/2}$ states. Note that there are four components of a $^4\Pi$ state, unlike the case of a ^4P state of an atom, which has only three components. In a first approximation (still considering the case of zero rotation) the energy of the multiplet components may be represented by the simple relation

$$T_e = T_0 + A\Lambda\Sigma, \tag{49}$$

where A is a constant characterizing the spin-orbit interaction [not to be confused with A of Eq. (43)]. Equation (49) means that in this approximation the multiplet components are equidistant, unlike the components of atomic multiplet states.

If we now introduce the effect of rotation on the spin splitting, we may distinguish several coupling cases, which were first discussed by Hund and are usually referred to as *Hund's coupling cases*. We shall briefly consider only the two that are most important.

In *Hund's case* (*a*) it is assumed that the spin-orbit coupling is large, while the coupling of the rotation of the nuclei with the electronic motion is very weak. In this case even in the rotating molecule the quantum number Ω remains a good quantum number. Figure 20 shows a vector diagram for the angular momenta in this case. The molecule is a symmetric top with $\boldsymbol{\Omega}$ instead of $\boldsymbol{\Lambda}$ as the angular momentum about the top axis. As a consequence, in Eq. (44) we have to replace Λ by Ω and note that the first rotational level for a given multiplet component is one with $J = \Omega$. Figure 21 shows, as examples, the rotational levels of a $^2\Pi$ and $^3\Delta$ state. In a first approximation, the two and three sets of rotational levels in the two electronic states are similar, except for the shift given by Eq. (49) and for the different number of missing levels at the bottom.

Hund's case (*b*) arises when the coupling of the spin with the internuclear axis is very weak and as a consequence the spin is coupled to the axis of rotation of the molecule. Such a situation almost always applies to Σ states; for small atomic numbers it also often applies to Π, Δ, . . . states. Figure 22 shows the vector diagram for this case: ***N*** is the total angular momentum apart

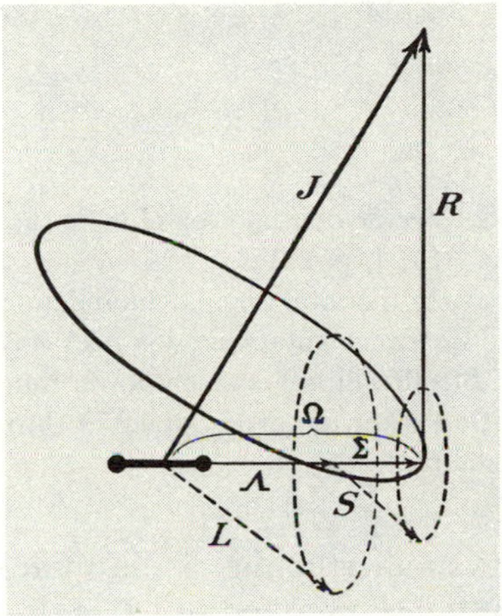

Fig. 20. Vector diagram for Hund's case (a) (from *MM* I, p. 219).

Only the total angular momentum ***J*** is fixed in space. The nutation of the figure axis about ***J*** is indicated by the solid-line ellipse; the precessions of ***L*** and ***S*** about the internuclear axis are assumed to be much faster (broken-line ellipses).

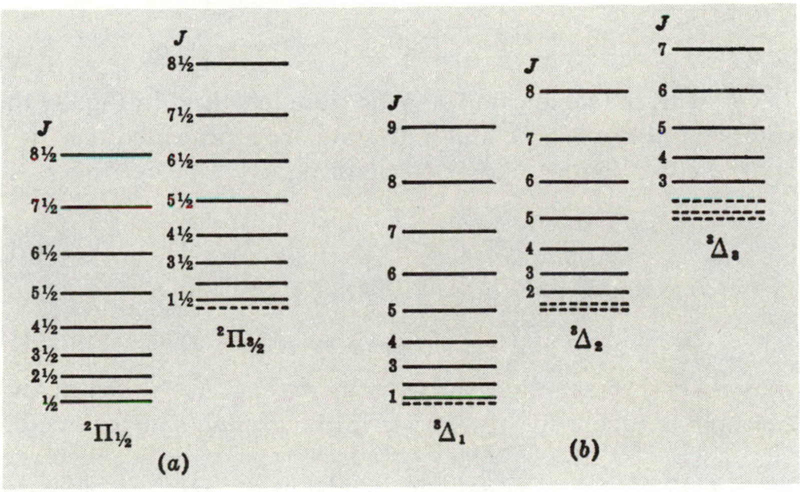

Fig. 21. The lowest rotational levels of a $^2\Pi$ and a $^3\Delta$ state in Hund's case (a) (from *MM* I, p. 220).

The dotted levels do not occur, since J must be $\geq \Omega$. The Λ-type doubling is ignored in this figure.

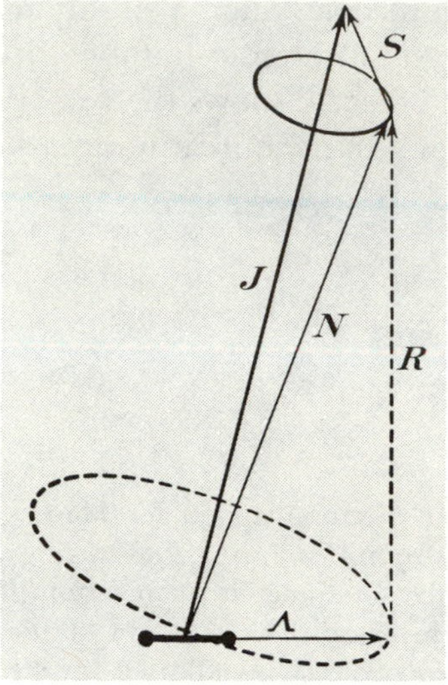

Fig. 22. Vector diagram for Hund's case (b) (from *MM* I, p. 221).

Again, only the total angular momentum $\boldsymbol{J}$ is fixed in space. The precession of $\boldsymbol{N}$ and $\boldsymbol{S}$ about $\boldsymbol{J}$ (solid ellipse) is much slower than the nutation of the figure axis about $\boldsymbol{N}$ (broken-line ellipse).

from spin[5]; its component is Λ. The smallest value of the quantum number N is Λ. The total angular momentum $\boldsymbol{J}$ of the molecule is the sum of $\boldsymbol{N}$ and $\boldsymbol{S}$. We have for the quantum numbers

$$J = N + S,\ N + S - 1,\ \ldots,\ |N - S|. \tag{50}$$

Levels with the same value of N lie close together. In Fig. 23 the rotational levels of a $^2\Sigma$ and a $^3\Sigma$ state are represented. For a $^2\Sigma$ state the two spin components are given by

$$\begin{aligned} F_1(N) &= B_vN(N+1) + \tfrac{1}{2}\gamma N, \\ F_2(N) &= B_vN(N+1) - \tfrac{1}{2}\gamma(N+1), \end{aligned} \tag{51}$$

or in other words the splitting is given by

$$\Delta\nu = F_1(N) - F_2(N) = \gamma(N + \tfrac{1}{2}); \tag{51a}$$

that is, it increases linearly with increasing N. A well-known example of such a splitting arises in the ground state of the CN

[5] In *MM*I the symbol K was used for what is now called N following a decision of the Joint Commission for Spectroscopy in 1952 (see J. Opt. Soc. Amer. *43*, 425, 1953). Some authors, unaware of this decision have continued to use K. The reason for the change is the possibility of confusion with the K used for symmetric top molecules (see Chap. IV, Section C1).

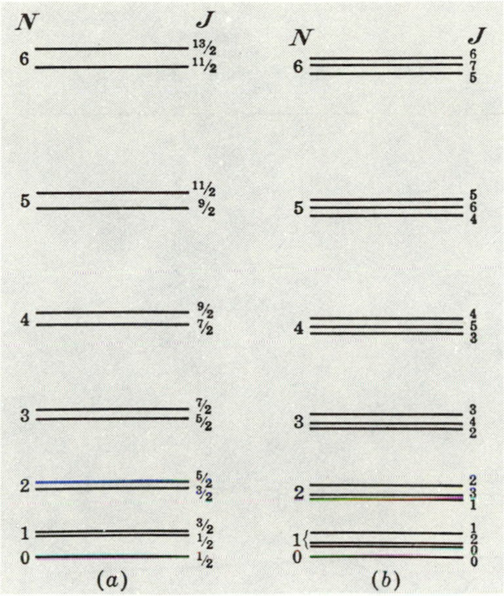

Fig. 23. The lowest rotational levels (a) of a $^2\Sigma$ and (b) of a $^3\Sigma$ state (from *MM* I, p. 222).

Usually the doublet or triplet splitting is much smaller relative to the rotational spacing than shown in this diagram.

free radical, for which the splitting constant γ has the value $+0.0076\ cm^{-1}$. For $^3\Sigma$ states the splitting formulae are rather more complicated and will not be given here (see *MM* I, p. 223). Instead, Fig. 24 shows the variation of the splitting with increasing N for the ground state of the SO free radical. The splitting starts from a non-zero value for $N = 1$ [for $N = 0$ there is only one level (see Fig. 23)]. For large N values the splitting varies approximately linearly with N as in $^2\Sigma$ states.

For multiplet states with $\Lambda \neq 0$ there is, in addition to the spin splitting, a Λ-doubling for each spin component. In Hund's case (a), for example, when the doublet splitting of a $^2\Pi$ state is large, the Λ-doubling in the $^2\Pi_{1/2}$ component varies with the first power of J, while in the $^2\Pi_{3/2}$ component it varies with J^3. In Fig. 25, as an example, the rotational levels of the ground states of OH and CH are given. The Λ-type doubling is exaggerated by a factor of 25. For CH the $^2\Pi$ state is "regular" ($^2\Pi_{1/2}$ is below $^2\Pi_{3/2}$), while for OH the $^2\Pi$ state is "inverted" ($^2\Pi_{3/2}$ is below $^2\Pi_{1/2}$). In both examples the doublet splitting is fairly small; that is, both are close to Hund's case (b).

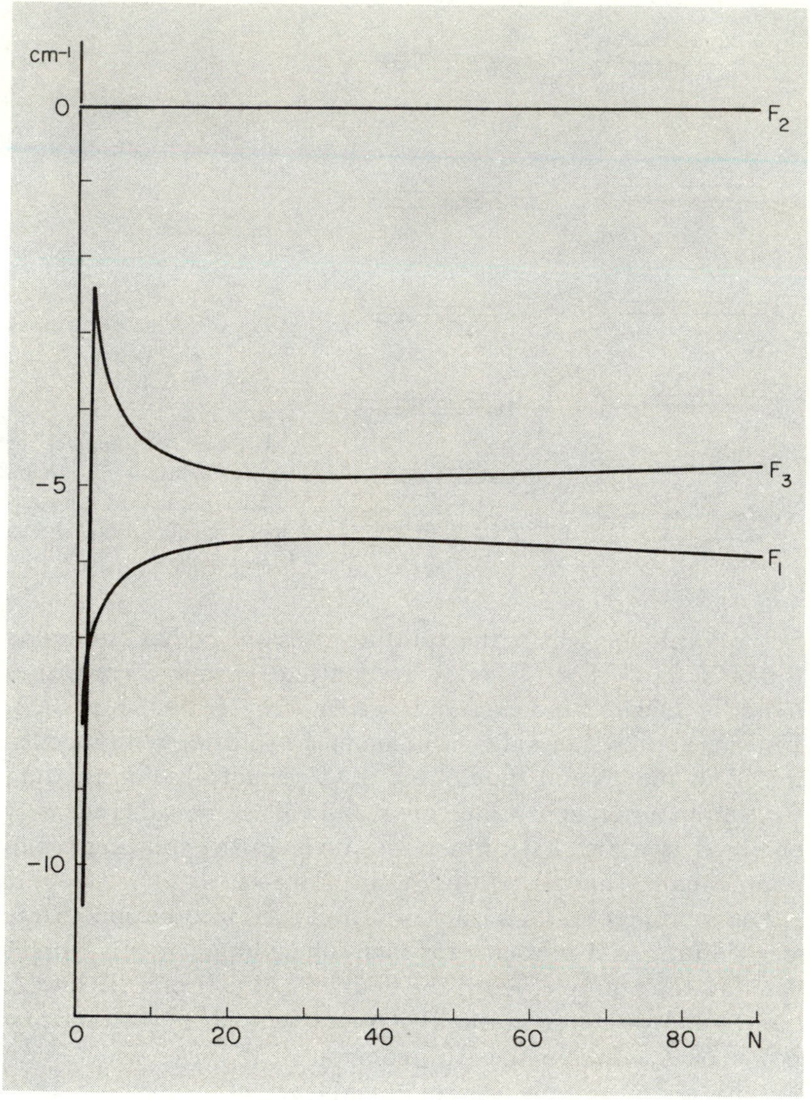

Fig. 24. Variation with N of the triplet splitting in the $^3\Sigma_g^-$ ground state ($v = 0$) of SO.

The energies of the three component levels after subtraction of $B_0N(N + 1)$ have been plotted as derived from Schlapp's formula (*MM* I, p. 223) using the splitting constants of Table 5. The observed energy levels for the lowest N values are plotted in Fig. 27.

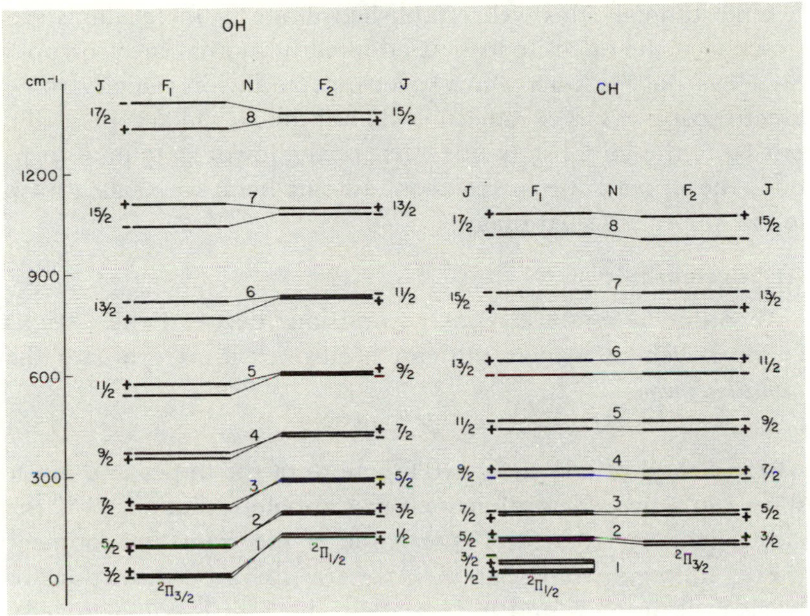

Fig. 25. Rotational levels in the ground states of OH and CH.

In both cases there is a rapid transition to case (b) as J increases. Levels with the same N in the two sets are connected by sloping lines. The Λ doubling is shown on a 25-fold enlarged scale. Note the reversal of the Λ doubling in the $^2\Pi_{1/2}$ component of OH between $N = 4$ and $N = 5$. According to Elitzur (145), the ± sign of the F_1 (½) level in CH should be changed to ∓.

B. TRANSITIONS, EXAMPLES

In designating transitions in diatomic and polyatomic molecular spectroscopy it has been established by usage and by decision of the International Joint Commission for Spectroscopy that the upper state should always come first irrespective of whether the transition is observed in absorption or emission, that is, in a transition A–B the symbol A refers to the upper, B to the lower state. If one wishes to indicate that the transition is observed in absorption or emission one writes $A \leftarrow B$ or $A \rightarrow B$ respectively. Similarly when one speaks about a 2–3 transition (or a 2–3 band or 2–3 line) one means that the implied quantum number (electronic, or vibrational, or rotational) has the value 2 in the upper and 3 in the lower state.

Unfortunately this well established usage in molecular spectroscopy is the opposite to that adopted in atomic spectroscopy. Some authors in microwave spectroscopy and in the electronic spectroscopy of larger molecules do not follow the accepted rule but write the initial state first, that is, the lower state in absorption, the upper state in emission. In this book we shall always follow the established usage.

(1) Selection rules

In order to establish which transitions between the various energy levels of a molecule can occur, we must evaluate the *transition moment*

$$R = \int \psi'^* M \psi'' \, d\tau, \tag{52}$$

where ψ' and ψ'' are the wave functions of the upper and lower state and where for ordinary electric dipole radiation M is the electric dipole moment. If the value of the transition moment (52) is different from zero, then the transition between the two levels considered is said to be "allowed"; if it is equal to zero, the transition is said to be "forbidden." The value of the integral in (52) depends on the quantum numbers that characterize the wave functions of the upper and lower states. The relations that must hold between the quantum numbers of the upper and lower states in order that the integral does not vanish and thus the transition is allowed are called *selection rules*. We must distinguish between rigorous electric dipole selection rules and approximate electric dipole selection rules. The former are independent of the degree of approximation introduced in the wave function; the latter are not.

As in the case of atoms (disregarding for the moment the effect of nuclear spin) we have for the total angular momentum $\boldsymbol{J}$ the selection rule

$$\Delta J = 0, \pm 1; \qquad J = 0 \not\leftrightarrow J = 0 \tag{53}$$

where $\Delta J = J' - J''$ is the change of J between upper and lower state and where $\not\leftrightarrow$ stands for "does not combine with." In addition we have the analogue of the Laporte rule for the parity, which is

$$+ \leftrightarrow -, \qquad + \not\leftrightarrow +, \qquad - \not\leftrightarrow -. \tag{54}$$

For homonuclear molecules we have, in addition, a selection rule for the symmetry property s and a:

$$s \leftrightarrow s, \qquad a \leftrightarrow a, \qquad s \not\leftrightarrow a. \tag{55}$$

For zero nuclear spin this rule holds not only for electric dipole radiation but for all other sorts of radiation and interactions with other molecules; it is an absolute rule which implies that if once only the s levels are present, no a levels will arise for all time. For nonzero nuclear spin the rule (55) is still very strong, but both s and a levels can now occur even though with different statistical weights (see p. 40).

When the nuclear spins of the two nuclei are not zero but $\boldsymbol{I}_1$ and $\boldsymbol{I}_2$, the total angular momentum is not $\boldsymbol{J}$ but

$$\boldsymbol{F} = \boldsymbol{J} + \boldsymbol{I}_1 + \boldsymbol{I}_2$$

and only the analogue of the rule (53)

$$\Delta F = 0, \pm 1, \qquad F = 0 \not\leftrightarrow F = 0 \tag{56}$$

holds rigorously for dipole radiation. But the rule (53) for the quantum number J of the total angular momentum apart from nuclear spin is still very strong (even though not rigorous), since the interaction with the nuclear spin is so weak. Irrespective of the presence or absence of nuclear spin, the parity rule (54) is rigorous for dipole radiation.

Since in a first approximation (see p. 39),

$$\psi = \psi_e \, \psi_v \, \psi_r, \tag{57}$$

there are, to the same approximation, separate selection rules for the electronic, the vibrational, and the rotational levels. For the electronic quantum numbers Λ and S we find the selection rules

$$\Delta\Lambda = 0, \pm 1 \tag{58}$$

and

$$\Delta S = 0. \tag{59}$$

The latter rule means that in this approximation *electronic states of different multiplicity do not combine with one another*. In addition, for Σ^+ and Σ^- electronic states the selection rule

$$\Sigma^+ \leftrightarrow \Sigma^+, \qquad \Sigma^- \leftrightarrow \Sigma^-, \qquad \Sigma^+ \not\leftrightarrow \Sigma^- \tag{60}$$

holds, and for homonuclear molecules for which the symmetry property g, u is defined the selection rule

$$g \leftrightarrow u, \qquad g \not\leftrightarrow g, \qquad u \not\leftrightarrow u \tag{61}$$

applies.

In Hund's case (a) the quantum number Σ is defined for which the selection rule

$$\Delta\Sigma = 0 \tag{62}$$

is found.

For the vibrational levels there is no strict selection rule, except that in the *vibration spectrum* transitions with

$$\Delta v = \pm 1 \tag{63}$$

have greatly preponderant intensity. For vibrational transitions between different *electronic* states, the Franck-Condon principle must be applied, which we shall discuss later.

For the combinations of the rotational levels we have, in addition to the rigorous rules (53), (54), and (55), in Hund's case (b) the selection rule

$$\Delta N = \pm 1, 0 \tag{64}$$

with the restriction that $\Delta N = 0$ *does not occur in* Σ–Σ *transitions.* In Hund's case (a) the selection rule (62) for Σ holds for the combinations of the sets of rotational levels of various multiplet components. The selection rules for N and Σ do not hold in coupling cases intermediate between (a) and (b).

As already mentioned, transitions that are not allowed by the electric dipole selection rules are called forbidden transitions. Such forbidden transitions may yet occur, either because of the possibility of radiation other than dipole radiation or because the selection rules are only approximately valid.

Just as classically an oscillating electric dipole produces radiation, so does an oscillating magnetic dipole or an oscillating electric quadrupole. However, *magnetic dipole and electric quadrupole radiations* are very much weaker than electric dipole radiation; consequently, especially in the study of free-radical spectra, they are not very important in an elementary treatment. It is only for the sake of completeness that we include here the selection rules for these types of radiation.

For magnetic dipole radiation, one finds the rigorous selection rules

$$\Delta J = 0, \pm 1; \qquad J = 0 \not\leftrightarrow J = 0; \tag{65}$$

$$+ \leftrightarrow +, \qquad - \leftrightarrow -, \qquad + \not\leftrightarrow -; \tag{66}$$

$$s \leftrightarrow s, \qquad a \leftrightarrow a, \qquad s \not\leftrightarrow a; \tag{67}$$

and the approximate selection rules

$$\Delta\Lambda = 0, \pm 1; \qquad \Delta S = 0; \tag{68}$$

$$\Sigma^+ \leftrightarrow \Sigma^-, \qquad \Sigma^+ \not\leftrightarrow \Sigma^+, \qquad \Sigma^- \not\leftrightarrow \Sigma^-; \tag{69}$$

$$g \leftrightarrow g, \qquad u \leftrightarrow u, \qquad g \not\leftrightarrow u. \tag{70}$$

For electric quadrupole radiation the rigorous selection rules are

$$\Delta J = 0, \pm 1, \pm 2; \tag{71}$$

$$J = 0 \not\leftrightarrow J = 0 \text{ or } 1, \qquad J = \tfrac{1}{2} \not\leftrightarrow J = \tfrac{1}{2};$$

$$+ \leftrightarrow +, \qquad - \leftrightarrow -, \qquad + \not\leftrightarrow -; \tag{72}$$

$$s \leftrightarrow s, \qquad a \leftrightarrow a, \qquad s \not\leftrightarrow a; \tag{73}$$

and the approximate rules

$$\Delta\Lambda = 0, \pm 1, \pm 2; \qquad \Delta S = 0; \tag{74}$$

$$\Sigma^+ \leftrightarrow \Sigma^+, \qquad \Sigma^- \leftrightarrow \Sigma^-, \qquad \Sigma^+ \not\leftrightarrow \Sigma^-; \tag{75}$$

$$g \leftrightarrow g, \qquad u \leftrightarrow u, \qquad g \not\leftrightarrow u. \tag{76}$$

If spin-orbit interaction is not negligible, transitions violating the rule $\Delta S = 0$ can occur as electric dipole radiation and do indeed form a fairly important group of forbidden transitions. Of these the singlet-triplet transitions are the most frequent.

In electric fields applied either externally or by collision with other molecules, transitions that are rigorously forbidden for electric dipole radiation in the free molecule may be induced to occur. This kind of radiation, *induced dipole radiation*, will not be of interest for free radicals.

(2) Rotation spectra and related spectra

If a molecule or radical has a permanent dipole moment, transitions between the rotational levels are allowed on the basis of the selection rules (53) and (54). If the ground electronic state is $^1\Sigma^+$, we see immediately from the rotational term-formula (10) that the wave numbers of the absorption or emission lines of this spectrum are given by

$$\nu = F(J+1) - F(J) = 2B(J+1) - 4D(J+1)^3 + \cdots. \quad (77)$$

Only a few free radicals have $^1\Sigma^+$ ground states, and no example of a pure rotation spectrum for such a case is known. However, in recent years several investigators have studied the rotation spectrum of the free radical SO, whose ground state is $^3\Sigma^-$. This spectrum lies in the microwave region. Figure 26 shows one of the lines as observed by Winnewisser, Sastry, Cook, and Gordy (140). The triplet splitting in the ground state is comparable to

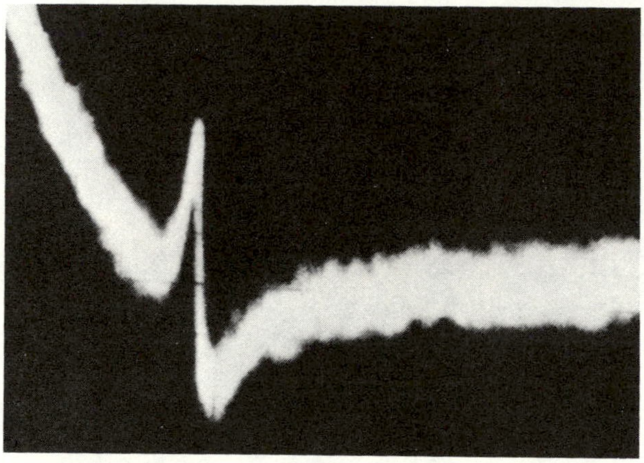

Fig. 26. Microwave rotation spectrum of the SO radical after Winnewisser, Sastry, Cook, and Gordy (140). The line corresponding to the transition $N = 4 \leftarrow 3$, $J = 3 \leftarrow 2$ is shown as it appears on the oscilloscope screen. Note that the deflection corresponds to the derivative of the signal; that is, at the center of the line the deflection goes through zero.

the separation of successive rotational levels as shown in Fig. 27. The transitions do not, therefore, form such a simple series as indicated by Eq. (77). Three groups of investigators under Lide (112), Gordy (139), and Morino (1) have obtained the transitions indicated in Fig. 27. The SO was produced in various ways, the best of which appears to be by reacting O atoms with OCS according to the reaction

$$O + OCS \longrightarrow SO + CO,$$

where the O atoms were obtained from an electric discharge.

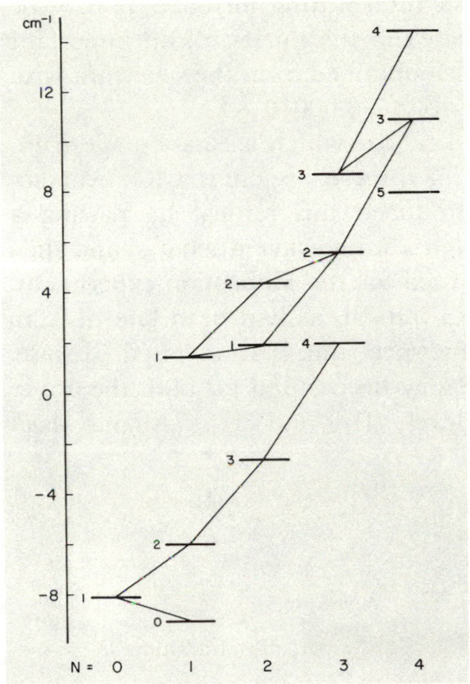

Fig. 27. Rotational energy-level diagram of the ground state of the SO radical.

The levels are grouped according to the *N* value. The transitions observed in absorption in the microwave region are indicated by the sloping lines.

Fitting the observed transitions to the splitting formula (see *MM* I, p. 223) yielded the B_0 and B_1 values given in Table 5.

Table 5. Molecular Constants of SO in its $^3\Sigma^-$ Ground State Obtained from the Microwave Spectrum

B_0	0.717949 cm^{-1}
B_1	0.712213
B_e	0.720817
α_e	0.005736
D_e	1.08×10^{-6}
β_e	0.09×10^{-6}
λ_0	5.27885
γ_0	−0.00562
r_e	1.48108 Å

The table also gives the rotational constants B_e and α_e, and the equilibrium internuclear distance r_e, as well as the coupling constants λ and γ that occur in the splitting formula. It may be noted that the separation between the nuclei, r_e, an important structural parameter, has been obtained from the spectrum with considerable accuracy (probably ±0.00010 Å).

Another diatomic free radical for which a rotation spectrum has been observed in the microwave region is ClO. Amano, Hirota, and Morino (2) produced this radical by passing a mixture of Cl_2 and O_2 through a microwave discharge and then into the wave guide that is used for the absorption experiment. The ground state of the ClO radical, as had been known from earlier studies of the electronic spectrum, is an inverted $^2\Pi$ state. Microwave transitions have now been found for both the upper and the lower component level, $^2\Pi_{1/2}$ and $^2\Pi_{3/2}$ [Amano *et al.*

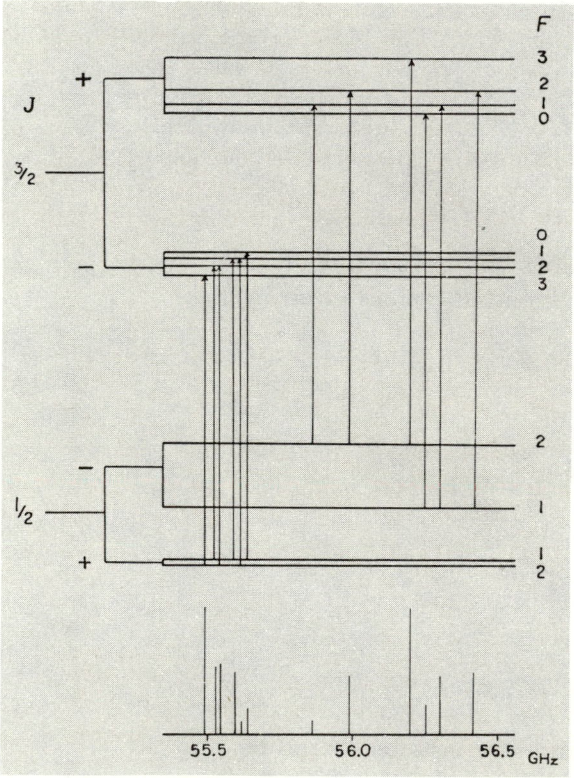

Fig. 28. Energy-level diagram for the rotation line $\frac{3}{2} \leftarrow \frac{1}{2}$ in the $^2\Pi_{1/2}$ ground state of ^{35}ClO, after Amano, Hirota, and Morino (2). [Copyright 1968 by Academic Press, Inc., New York.]

At the left the levels before introduction of Λ-type doubling and hyperfine structure are shown; a little to the right is shown the effect of Λ-type doubling, while the main part of the diagram shows the full observed splitting. At the bottom the observed spectrum is plotted, the height of the vertical lines indicating the intensities.

(3)]. In Fig. 28 the observed transitions in the $^2\Pi_{1/2}$ state of ^{35}ClO are shown in an energy-level diagram for the $\frac{3}{2} \leftarrow \frac{1}{2}$ transition. Each level is double on account of Λ-type doubling; therefore, according to the selection rule (54), two components would be expected for the transition $J = \frac{3}{2} \leftarrow J = \frac{1}{2}$ if the effect of nuclear spin is neglected. Actually the coupling of the nuclear spin of Cl ($I = \frac{3}{2}$) with J produces two sublevels for each of the levels with $J = \frac{1}{2}$ and four sublevels for each of the levels with $J = \frac{3}{2}$. These sublevels are numbered by the quantum number F (see p. 51). Applying the selection rule (56) for F, we see that six component lines are expected for each of the two Λ components, and they have indeed been observed. The resulting molecular constants are (for ^{35}ClO)

$$
\begin{aligned}
B_0 &= 18602.9 \text{ Mc/sec} = 0.620524 \text{ cm}^{-1}, \\
B_e &= 18694 \text{ Mc/sec} = 0.62356 \text{ cm}^{-1}, \\
r_e &= 1.569 \text{ Å}.
\end{aligned}
$$

Since in this case no transitions for $v = 1$ have been observed, the value of B_e was derived indirectly from the difference in B_0 for the two isotopic molecules ^{35}ClO and ^{37}ClO (for a discussion of the rotational isotope effect see MMI, p. 143f.).

In a heteronuclear molecule the transition between the two Λ-components of given J is electric-dipole allowed. The first and so far only diatomic example of a spectrum of this type was observed by Townes and his collaborators (37)(40) for OH. For this radical they observed a number of such *Λ-doubling transitions*, the most important being the one for the lowest rotational level $J = \frac{3}{2}$ of the $^2\Pi_{3/2}$ ground state (see Fig. 25). Again a complication arises on account of the nuclear spin of the proton ($I = \frac{1}{2}$). As shown in Fig. 29, there are four components of this transition, which without nuclear spin would coincide. The two components with $\Delta F = 0$ are the strongest. The average of their frequencies gives a precise value for the Λ-doubling, but no information about the value of B_0 is obtained. By studying the Stark effect (see *MM* I, p. 307) of the Λ-doubling lines, Powell and Lide (113) have obtained a very precise value for the electric

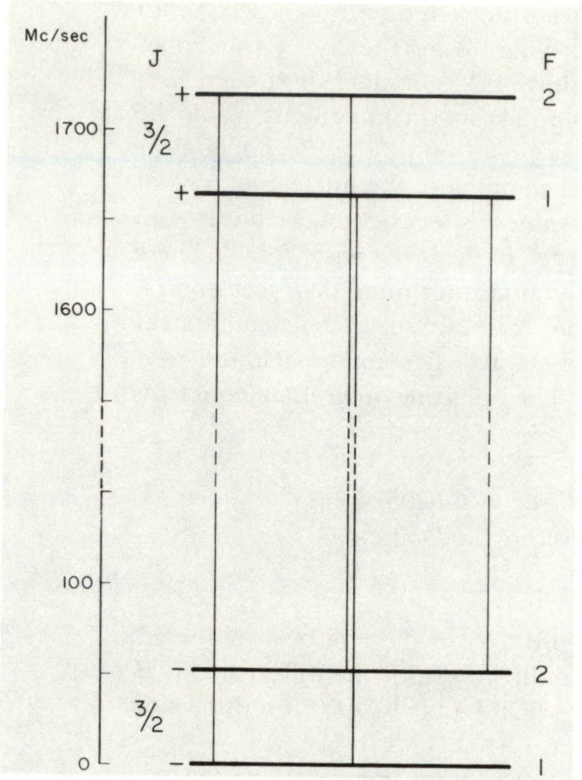

Fig. 29. Energy-level diagram showing the transitions of the Λ-doubling spectrum of OH.

The figure is drawn to scale for the $J = \frac{3}{2}$ level of the $^2\Pi_{3/2}$ state.

dipole moment of the OH radical—namely, $\mu = 1.660 \pm 0.010$ deb.[6]

Since 1963 the Λ-doubling spectrum of OH for $J = \frac{3}{2}$ has become of great importance in astronomy, since it has been observed by radio telescopes in both absorption and emission in the *interstellar medium* and even, quite recently, in the atmospheres of a few stars. In some of these astronomical objects a strange anomaly in the intensity ratios of the four lines (see Fig. 29) has been found. In addition, a few pointlike sources have been observed that emit the Λ-doubling lines of OH with considerable intensity. These sources cannot be identified with any

[6] Previously Madden and Benedict (88) had derived an approximate value of 1.54 deb from observations of the pure rotation spectrum in the far infrared.

known stars. Their pointlike character, together with the polarization properties of the radiation, suggests strongly that we have here the effects of a cosmic maser, although the exact nature of the process by which an inverted population is produced is not yet clear. It seems likely that these maserlike objects represent proto-stars—that is, stars in formation.

Spectra similar to that of OH are expected for the radicals CH, SiH, and SH, but despite considerable searching they have not yet been found. The electron-spin resonance spectrum of SH has been observed by Radford and Linzer (116), who on the basis of this spectrum have predicted the exact positions of the Λ-doubling lines.

Another spectrum in the radio-frequency region that must be mentioned here is the *spin-reorientation spectrum:* it consists of transitions between the various spin components of a $^2\Sigma$ or $^3\Sigma$ state without change of the rotational quantum number N. These transitions are electric-dipole forbidden, since the levels involved have the same parity, but they are magnetic-dipole allowed. Such transitions were first observed for the O_2 molecule and more recently have been studied by Jefferts (76) for the H_2^+ molecular ion. We shall consider only the latter spectrum here. H_2^+ has a $^2\Sigma_g^+$ ground state. The rotational levels are split according to the previous formula (51a). The magnitude of the splitting constant γ has been predicted from theory to be 0.00153 cm^{-1}. Here again the situation is complicated by the nuclear spin. However, for the even rotational levels, since the nuclear spins are antiparallel—that is, the total nuclear spin is zero—no such complication arises. Jefferts observed for $N = 2$ two radio-frequency lines which, according to his estimate, correspond to the vibrational levels $v = 5 \pm 1$ and $v = 6 \pm 1$. The frequencies of the observed lines, which are 75.598 and 70.231 Mc/sec, according to Eq. (51a) represent directly the value of 2.5 γ. Luke (87) in our laboratory has recently calculated from first principles the dependence of γ on the vibrational quantum number v. With a correction by Somerville (122) this calculation gives

$$\Delta\nu_{v=5} = 75.5, \qquad \Delta\nu_{v=6} = 70.1, \qquad \Delta\nu_{v=7} = 64.8 \quad \text{Mc/sec.}$$

Apparently the two lines observed by Jefferts do correspond to $v = 5$ and $v = 6$.

For $N = 1$ the nuclear spin splitting must be considered. Indeed it is much larger than the splitting produced by the coupling of the electron spin and rotation. The corresponding spectrum is in the region of the 21-cm line of atomic hydrogen. This spectrum, also, has been observed by Jefferts for $v = 4, 5, \ldots 8$. The observed wavenumbers agree very well with the predictions by Luke and Somerville. A search for the corresponding H_2^+ lines with $v = 0$ in the interstellar medium by radiotelescopes has not yet been successful [Jefferts et al. (76a)].

(3) Rotation-vibration spectra

If a transition from one vibrational level to another of a given electronic state (usually the ground state) occurs, we obtain a *vibration spectrum*, and if we include in our considerations the simultaneous change of rotational level we obtain a rotation-vibration spectrum. The various possible rotational transitions for a given vibrational transition form a *band*—more particularly, a *rotation-vibration band.* The wave numbers of the lines of such a band are the sum of a vibrational and a rotational contribution. The vibrational part is given by

$$\nu_v = G(v') - G(v''), \tag{78}$$

where v' and v'' are the vibrational quantum numbers of upper and lower state. In absorption at low temperature $v'' = 0$, and we have

$$\begin{aligned} \nu_v(\text{abs}) &= G(v') - G(0) = G_0(v') \\ &= \omega_0 v' - \omega_0 x_0 v'^2 + \cdots. \end{aligned} \tag{79}$$

The rotational part, assuming a $^1\Sigma$ state, is given by

$$\nu_r = F_{v'}(J') - F_{v''}(J''). \tag{80}$$

Substituting $F_v(J)$ from Eq. (24) but neglecting the small centrifugal stretching term and combining vibrational and rotational contributions, we obtain (omitting the subscripts v of B' and B'')

$$\nu = \nu_v + B'J'(J'+1) - B''J''(J''+1), \tag{81}$$

where B' and B'' are the rotational constants in the upper and lower states and J' and J'' are the corresponding J values. Usually, for rotation-vibration spectra one writes $\nu_v = \nu_0$. This wave number represents the *origin of the band*—that is, the wave number corresponding to $J' = J'' = 0$.

For a Σ state the selection rule (53) [with the restriction that $\Delta J = 0$ does not occur] requires that

$$J' = J'' \pm 1.$$

Substituting into (81) and writing, in accordance with common practice, ν_0 in place of ν_v, J in place of J'', we obtain for $J' = J + 1$

$$\begin{aligned}\nu_R &= \nu_0 + B'(J+1)(J+2) - B''J(J+1) \\ &= \nu_0 + 2B' + (3B' - B'')J + (B' - B'')J^2 \end{aligned} \qquad (82)$$

and for $J' = J - 1$

$$\begin{aligned}\nu_P &= \nu_0 + B'(J-1)J - B''J(J+1) \\ &= \nu_0 - (B' + B'')J + (B' - B'')J^2. \end{aligned} \qquad (83)$$

These two equations represent two series of lines called R branch ($\Delta J = +1$) and P branch ($\Delta J = -1$). Figure 30 shows the corresponding transitions in an energy-level diagram. Note that the first line of the R branch has $J = 0$, that of the P branch has $J = 1$. The two branches can also be represented by a single formula

$$\nu = \nu_0 + (B' + B'')m + (B' - B'')m^2, \qquad (84)$$

where $m = J + 1$ for the R branch and $m = -J$ for the P branch. In other words, we have a single series of lines in which one line is missing (at ν_0), forming the so-called *zero gap*.

If the ground state of the molecule is a Π state, $\Delta J = 0$ as well as $\Delta J = \pm 1$ is possible. From Eq. (81) we then obtain, in addition to the R and P branches, a third branch, the Q branch, which is given by

$$\nu_Q = \nu_0 + (B' - B'')J + (B' - B'')J^2. \qquad (85)$$

Since B' and B'' are not very different (they differ only by a small integral multiple of α_e), all lines of the Q branch lie close to ν_0. Because $J = 0$ does not exist in a Π state, the first lines in

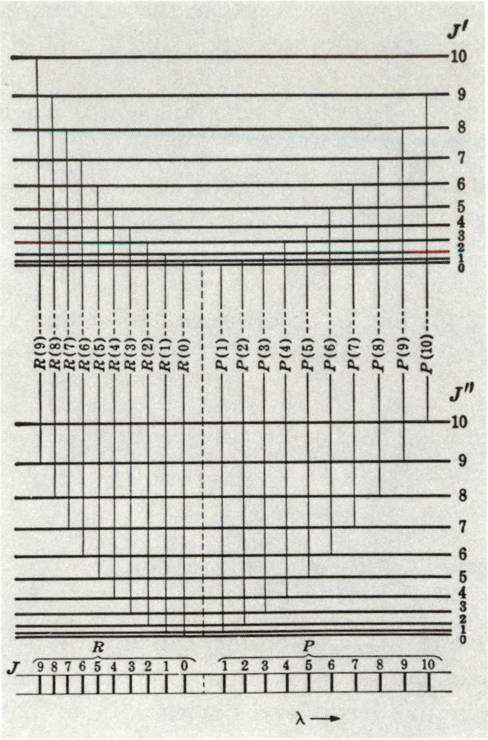

Fig. 30. Energy-level diagram explaining the fine structure of a rotation-vibration band (from *MM* I, p. 112).

At the bottom the resulting spectrum is shown schematically.

both the Q and R branches are those with $J = 1$, while the first line in the P branch is that with $J = 2$ (that is, $J' = 1$). The gap in the P, R series is thus increased compared to the rotation-vibration spectrum of a molecule in a Σ state, but the Q branch lies in this gap.

Up to now the rotation-vibration spectrum of only one diatomic free radical has been observed in the gaseous phase: that of OH. In 1950 Meinel (91) observed for the first time with fairly high resolution the spectrum of the night sky in the photographic infrared and found a new group of bands, reproduced in Fig. 31. Although Meinel first thought that this was a new electronic band system of OH, it was soon brought out in correspondence with him that this is part of the rotation-vibration spectrum of the OH molecule. Because of the doublet nature of the ground

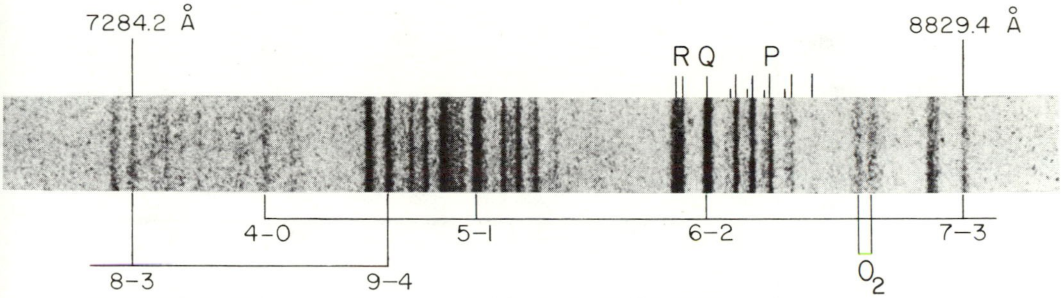

Fig. 31. Rotation-vibration spectrum of OH as observed by Meinel (91) in the light of the night sky.

state ($^2\Pi$), each band consists of two subbands, of which the one corresponding to $^2\Pi_{3/2}$ is by far the stronger. Each subband has three branches: P, Q, and R. The Q branch forms an unresolved line near the center of the band; it has appreciable intensity only for low J values.

By the extension of Meinel's spectra to longer wavelengths, five sequences have now been observed [Chamberlain and Roesler (16), Gush (see 133)]:

6–0, 7–1, . . . , 9–3,
5–0, 6–1, . . . , 9–4,
4–0, 5–1, . . . , 9–5,
3–0, 4–1, . . . , 9–6,
2–0, 3–1, . . . , 9–7.

These are represented in the vibrational energy-level diagram, Fig. 32. All sequences break off at $v' = 9$, even though the last band is a very strong one. This breaking off has been explained [Bates and Nicolet (6), Herzberg (55)] by the assumption that OH emission in the upper atmosphere arises by a reaction between hydrogen atoms and ozone molecules according to the following scheme:

$$H + O_3 \longrightarrow OH + O_2. \qquad (86)$$

It can easily be verified that the heat of this reaction is 3.3_0 eV, which corresponds exactly to the energy of the $v = 9$ level above $v = 0$. If it is further assumed that reaction (86) is followed by

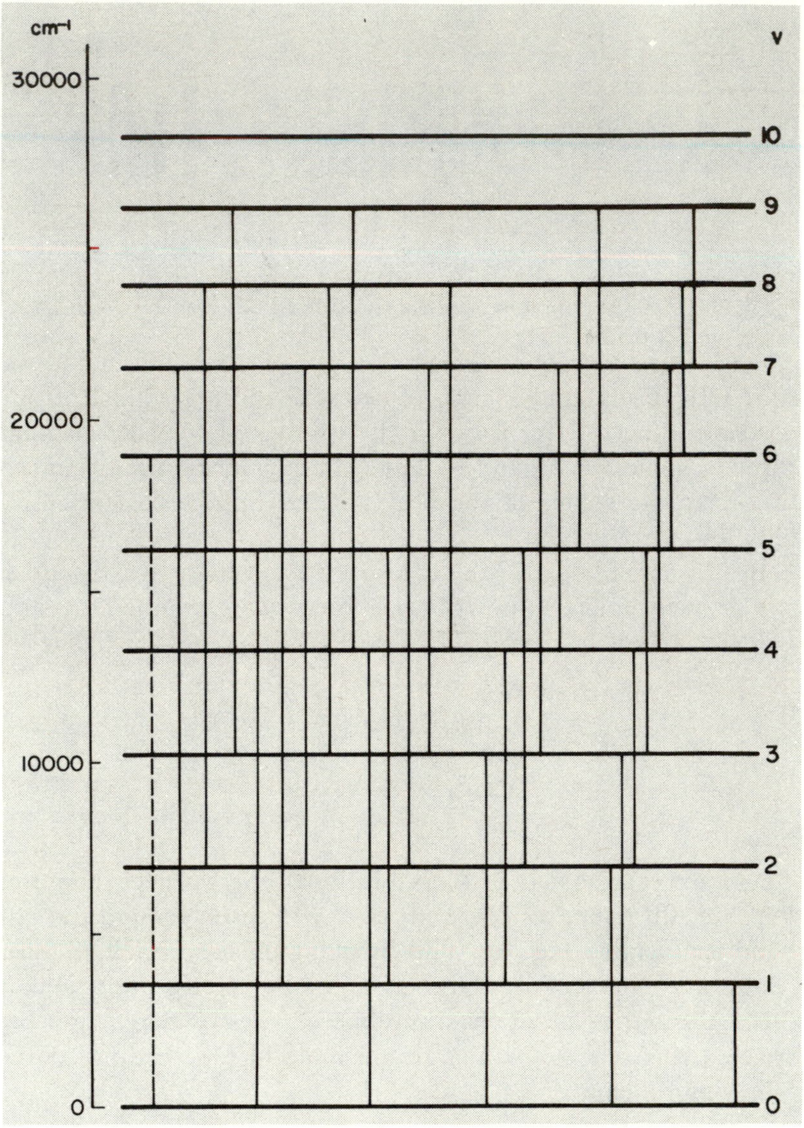

Fig. 32. Vibrational energy-level diagram of OH in its ground state showing the transitions observed in the night sky and in the $H + O_3$ flame.

The transitions with the same Δv form close-lying groups called sequences (see p. 67). Additional members of the $\Delta v = 1$ sequence have been observed by MacDonald, Buijs and Gush (87a).

$$OH + O \longrightarrow O_2 + H + 0.73 \text{ eV}, \qquad (87)$$

one has a chain reaction in which a single H atom can produce many excitations of OH, causing at the same time the destruction of O_3 and the recombination of O atoms to O_2 molecules.

While originally the reaction (86) represented an *ad hoc* assumption in order to account for the observed emission, a few years later McKinley and Garvin (90)(42) observed the chemiluminescence produced by the reaction between hydrogen atoms and ozone molecules in the laboratory and found it to have essentially the same spectrum, also breaking off at $v = 9$. Since that time the rotation-vibration spectrum of OH has also been observed in oxyhydrocarbon flames by Herman and Hornbeck (51) and extended to longer wavelengths. This work has yielded the most precise rotational constants of OH in its ground state, while the best vibrational constants have been obtained by a combination of laboratory and night-sky data by Chamberlain and Roesler (16). These constants are summarized in Table 6.

Table 6. Rotational and Vibrational Constants in the Ground State of the OH Radical

B_e	18.867 cm^{-1}
α_e	0.708
γ_e	0.00207
ω_e	3739.94
$\omega_e x_e$	86.350
$\omega_e y_e$	+0.9046
$\omega_e z_e$	+0.05763
r_e	0.9707$_8$ Å

γ_e is the coefficient of $(v + \frac{1}{2})^2$ in the expansion (21)

(4) Electronic spectra

Vibrational structure. In the general case all three forms of energy (electronic, vibrational, and rotational) change. Let us first consider the vibrational structure; that is, let us neglect rotation and consider all possible vibrational transitions for a given electronic transition. Starting with a given upper (or lower) vibrational level, series of transitions to all the vibrational levels of the lower (or upper) state are possible; they are called v'' *progressions* (or v' *progressions*). In Fig. 33 such progressions are shown in an energy-level diagram. We obtain all vibrational transitions in a band system by considering either all possible v'' progressions [Fig. 33(a)] or all possible v' progressions [Fig. 33(b)]. Which of these transitions are strong and which are weak will be discussed further below (p. 69f.).

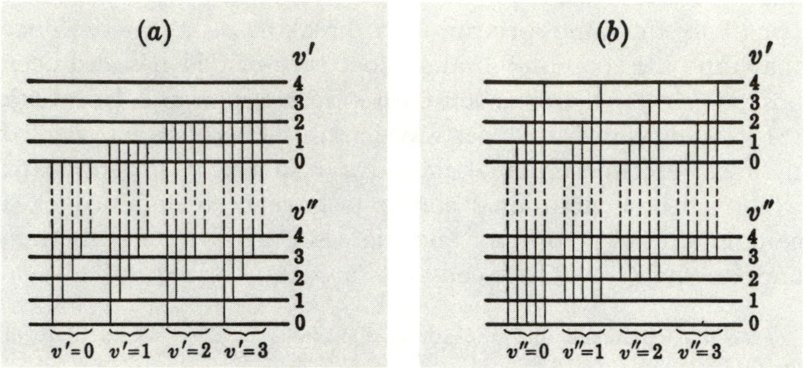

Fig. 33. Energy-level diagrams representing progressions of bands: (a) v''-progressions, (b) v'-progressions (from *MM* I, p. 153).
The individual progressions are indicated by braces.

The general expression for the wave numbers of the lines in an electronic transition consists of three contributions—electronic, vibrational, and rotational:

$$\nu = \nu_e + \nu_v + \nu_r. \tag{88}$$

ν_e, the origin of the band system, is fixed for a given electronic transition and is given by the difference of the purely electronic energies—that is, the energies of the minima of the potential functions of the two states

$$\nu_e = T_e' - T_e''. \tag{89}$$

The contributions ν_v and ν_r are similar to those in the rotation-vibration spectrum,

$$\nu_v = G'(v') - G''(v''), \tag{90}$$

$$\nu_r = F'(J') - F''(J''), \tag{91}$$

except that now G' and F' belong to a different electronic state from G'' and F'' and the vibrational and rotational constants entering into them may have quite different values.

Neglecting for the time being ν_r, which in general is small compared to ν_v, and using only the first two terms of Eq. (18) for $G(v)$, we obtain for the bands of a band system

$$\nu = \nu_e + \omega_e'(v' + \tfrac{1}{2}) - \omega_e'x_e'(v' + \tfrac{1}{2})^2 - [\omega_e''(v'' + \tfrac{1}{2}) - \omega_e''x_e''(v'' + \tfrac{1}{2})^2]. \tag{92}$$

For a v' progression (with fixed v'') we obtain

$$\nu = [\nu_e - G''(v'')] + \omega_e'(v' + \tfrac{1}{2}) - \omega_e'x_e'(v' + \tfrac{1}{2})^2, \tag{93}$$

where the term in square brackets represents the energy difference between the potential minimum of the upper state and the (fixed) lower vibrational level. Instead of (93) one often writes

$$\nu = \nu(0, v'') + \omega_0'v' - \omega_0'x_0'v'^2, \tag{94}$$

where $\nu(0, v'')$ is the wave number of the first band ($v' = 0$) of the progression and where ω_0 and ω_0x_0 are the vibrational constants defined by Eq. (18a). Similarly for a v'' progression ($v' =$ constant) we obtain

$$\nu = \nu(v', 0) - (\omega_0''v'' - \omega_0''x_0''v''^2). \tag{95}$$

Since ω_0x_0 is small compared to ω_0, we see that a progression is a series of bands with only slowly changing separations, the magnitudes of which represent directly the vibrational intervals ΔG in the upper or lower state (see Fig. 33).

If the vibrational frequency ω in the upper state is not very different from that in the lower state, it is immediately obvious from Fig. 34 and Eq. (92) that transitions with the same change Δv ($= v' - v''$) of the vibrational quantum number lie fairly close together. These groups of transitions are called *sequences*.

In Fig. 35, as an example, the near-ultraviolet emission spectrum of the PN radical is shown. The assignments of the bands are indicated below the spectrum. The v'' progressions are marked in separate lines; the sequences are clearly visible. The vibrational analysis of such a band system is very simple: the

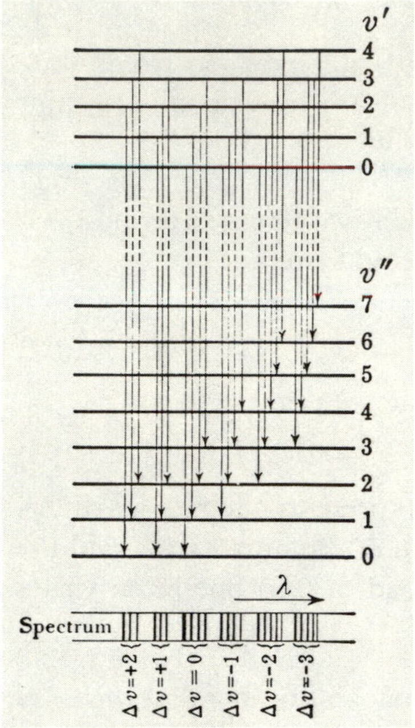

Fig. 34. Energy-level diagram representing sequences of bands (from *MM* I, p. 159).

The figure is drawn roughly to scale for the PN bands shown in the spectrogram, Fig. 35.

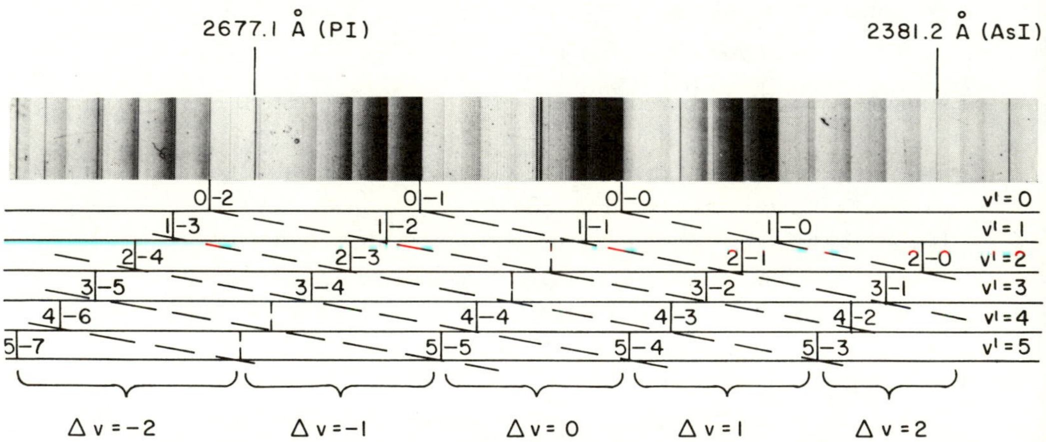

Fig. 35. Spectrum of the PN molecule in emission [after Curry, Herzberg, and Herzberg (25), from *MM* I, p. 33].

The v''-progressions and the sequences are indicated below. The broken leading lines refer to bands that have not been observed.

0–0 band can easily be identified since it is the starting point of the first v' progression and the first v'' progression, which have somewhat different spacings corresponding to the ΔG values of the upper and lower state. Conversely these ΔG values can be determined in this way.

In order to check the vibrational analysis of a band system and to present the data, we set the wave numbers out in a *Deslandres table*. Such a table for PN is given in Table 7. In horizontal rows are the successive v'' progressions, in vertical rows the v' progressions. If the vibrational analysis is correct, the wave-number differences of corresponding bands in different v'' progressions must be constant. The constant wave-number difference of the first two v'' progressions corresponds to $\Delta G'(\frac{1}{2})$. The constant difference of the next two is $\Delta G'(\frac{3}{2})$, and so on. Similarly, the constant wave-number difference of the first two v' progressions gives $\Delta G''(\frac{1}{2})$, and so on. In Table 7 the wave numbers refer to band heads, and so the accuracy of the agreement of these combination differences is only moderately good. If band origins were used instead, exact agreement within the accuracy of the measurements would be expected.

The intensity distribution in the progressions varies greatly. Figure 36 shows three cases for a v' progression with $v'' = 0$—that is, for the progression that would occur in absorption at low temperature. In the first case, the intensity decreases very rapidly from the first band on; in the second case the intensity first increases to a maximum and then decreases; and in the third case the intensity is very small for low v' values and only gradually attains larger values, eventually reaching a maximum that may or may not lie on the long-wavelength side of the convergence limit. Beyond the convergence limit there is a continuous spectrum corresponding to dissociation (see Chapter V).

An explanation for the different intensity distributions is provided by the *Franck-Condon principle*. According to the original semiclassical idea of Franck, the "quantum jump" from one electronic state to the other is very fast compared to the motion of the nuclei; therefore, immediately after the quantum jump, the nuclei still have the same position and velocity as before it.

Table 7. Deslandres Table of the PN Bands

v' \ v''	0		1		2		3		4		5		6		7		8		9		10
0	39698.8	1322.3	38376.5	1307.8	37068.7		......		......		......		......		......		......		......		......
	1087.4		1090.7		1086.8																
1	40786.2	1319.0	39467.2	1311.7	38155.5	1294.2	36861.3		......		......		......		......		......		......		......
	1072.9		1069.0				1071.6														
2	41859.1	1322.9	40536.2		......		37932.9	1280.4	36652.5	1265.3	35387.2		......		......		......		......		......
			1061.2						1060.0		1059.2										
3	......		41597.4	1309.1	40288.3		......		37712.5	1266.1	36446.4	1252.4	35194.0		......		......		......		......
					1042.9				1043.9				1042.6								
4	......		......		41331.2		......		38756.4		......		36236.6	1238.3	34998.3		......		......		......
															1029.4						
5	......		......		......		41066.1		......		38519.4		......		36027.7	1225.5	34802.2		......		......
							1015.9														
6	......		......		......		42082.0		......		......		......		......		......		34607.1	1194.5	33412.6
																					998.7
7	......		......		......		......		41798.3		......		......		......		......		......		34411.3
8	......		......		......		......		......		41522.6		......		......		......		......		......
9	......		......		......		......		......		......		41239.4		......		......		......		......

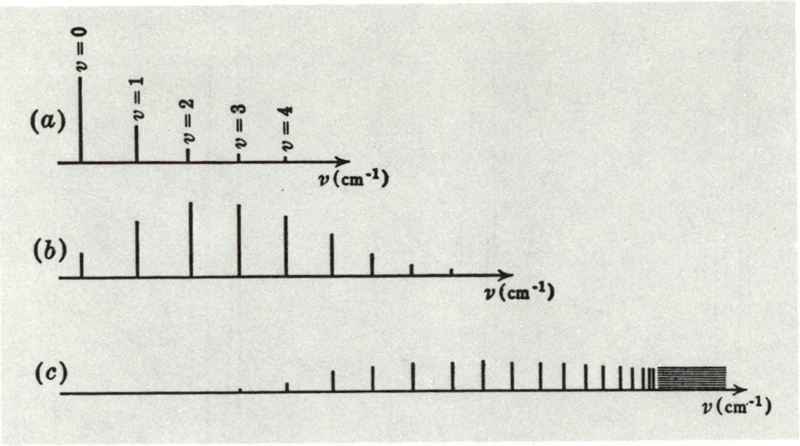

Fig. 36. Typical cases of intensity distribution in a v'-progression in absorption ($v'' = 0$) (from *MM* I, p. 194).

On this basis the potential diagrams in Fig. 37 explain readily the three cases of intensity distribution given in Fig. 36. In absorption the molecule is initially at the minimum of the lower potential curve if we disregard zero-point vibration. The quantum jump produced by the absorption of a light quantum brings the molecule to a point of the upper potential curve with the same r and (approximately) zero velocity. In Fig. 37(b), for example, that is the point B. Because of the change in potential function the molecule now starts oscillating between B and C. The transition from the ground state ($v'' = 0$) to a level in the neighbourhood of BC will therefore be the most probable transition, explaining the intensity distribution in Fig. 36(b). The two other cases (a) and (c) are similar.

Condon developed the wave-mechanical formulation of this principle by starting out from the Born-Oppenheimer approximation [see Eqs. (37) and (41)]. In this approximation, neglecting rotation, one immediately obtains for the transition moment (52) the expression

$$R_{e'v'e''v''} = \int \psi_{ev}'^{*} M \psi_{ev}'' \, d\tau_{ev} = \int \psi_e'^{*} \psi_v'^{*} M \psi_e'' \psi_v'' \, d\tau_e \, d\tau_v. \tag{96}$$

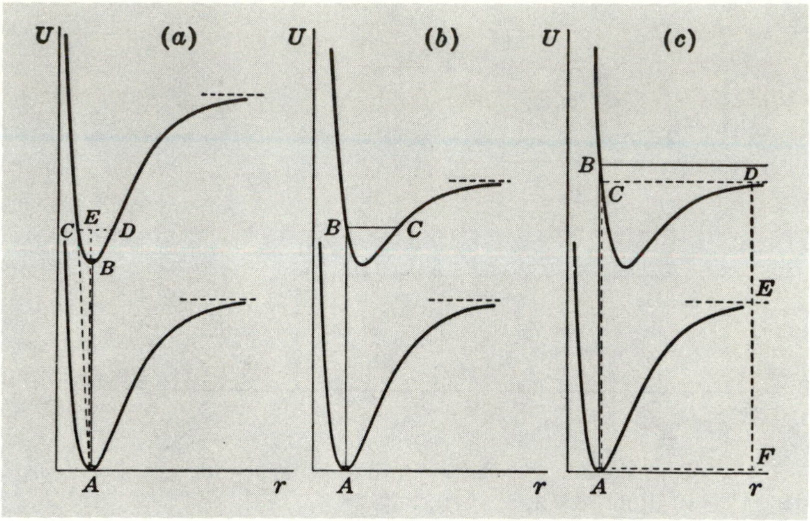

Fig. 37. Potential curves explaining the three cases of intensity distribution in absorption (Fig. 36) on the basis of the Franck-Condon principle (from *MM* I, p. 195).

In (a) in order to reach the level *CD* there would either have to be a sudden change of position $E \rightarrow C$ or a sudden change *EB* of the kinetic energy of the nuclei, both of which contradict the Franck-Condon principle; therefore in this case the transition $A \rightarrow B$ is the strongest. In (c) *AC* gives the energy of the dissociation limit, *EF* the dissociation energy of the ground state, and *DE* the excitation energy of the dissociation products upon photodissociation.

The dipole moment M can be resolved into an electronic and a nuclear component as follows:

$$M = M_e + M_n. \quad (97)$$

Substituting into (96) gives

$$R_{e'v'e''v''} = \int \psi_v'^* \psi_v'' \, dr \int \psi_e'^* M_e \psi_e'' \, d\tau_e + \int \psi_v'^* M_n \psi_v'' \, dr \int \psi_e'^* \psi_e'' \, d\tau_e. \quad (98)$$

For each r value the integral $\int \psi_e'^* \psi_e'' \, d\tau_e$ vanishes, since the ψ_e are orthogonal to one another and therefore Eq. (98) simplifies to

$$R_{e'v'e''v''} = R_{e'e''} \int \psi_v'^* \psi_v'' \, dr, \quad (99)$$

where

$$R_{e'e''} = \int \psi_e'^* M_e \psi_e'' \, d\tau_e \tag{100}$$

is the electronic transition moment, which in the Born-Oppenheimer approximation is independent of r.

Remembering that the intensities are proportional to the squares of the transition moments, we see from Eq. (99) that in a first approximation the relative intensities of the bands in a band system are given by the squares of the corresponding overlap integrals

$$\left[\int \psi_v'^* \psi_v'' \, dr\right]^2 \tag{101}$$

of the vibrational eigenfunctions. Figure 38 shows for $v' = 0, 1, 2, 4$ and $v'' = 0$ the vibrational eigenfunctions for a case similar to (b) of Figs. 36 and 37. Qualitatively it can be seen from this figure that the overlap integral (101) reaches a maximum for $v = 2$. It will have smaller but nonzero values on either side of this maximum, showing the effect of the wave nature of matter on the original semiclassical principle of Franck.

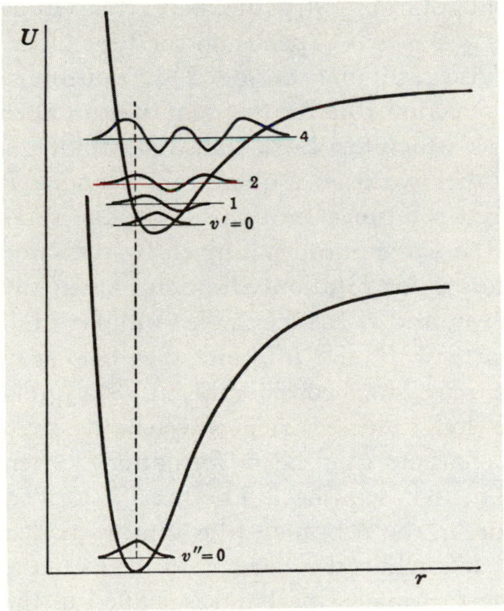

Fig. 38. Potential curves and vibrational eigenfunctions explaining the wave-mechanical form of the Franck-Condon principle.

The potential functions of upper and lower states are assumed to have such a relative position that the "best" overlapping of the vibrational eigenfunctions occurs for $v' = 2$, $v'' = 0$ (see the broken vertical line).

It is clear from Fig. 38 that if one considers emission from a state with $v' = 2$, for example, there will be two maxima in the intensity distribution of the v'' progression with this upper state, and similarly for other v' values (except $v' = 0$). As a result one finds for the intensity distribution in the Deslandres table a parabolic curve which is well illustrated by Table 7: it is called the *Condon parabola*. A few fairly intense bands off this parabola can readily be explained on the basis of wave mechanics as due to accidentally favorable overlap of the two vibrational wave functions.

As an example of a progression observed in absorption we give in Fig. 39 the absorption spectrum of the SO radical recently reinvestigated by Colin (21), which shows the convergence of the progression very clearly. It has been observed up to the last band before the limit. This limit occurs at 52,500 cm^{-1} (1904.8 Å) and corresponds to the energy required to dissociate the SO radical into an S atom in the ^1D state and an O atom in its ^3P ground state (see Chapter V).

Rotational structure. The rotational structure of a given vibrational transition—that is, of a *band*—depends on the type of the electronic states involved. We shall first consider Σ–Σ transitions. For these transitions the selection rule for the quantum number N is $\Delta N = \pm 1$ (see p. 52), which for $^1\Sigma$–$^1\Sigma$ transitions is identical with $\Delta J = \pm 1$. In other words, we obtain an R and a P branch, just as for infrared rotation-vibration bands; the rotational contribution ν_r to the wave number is given by the same formulae as already discussed for rotation-vibration bands: the formulae (82) and (83) for ν_R and ν_P [or the single formula (84)]. The only difference is that now B' and B'', since they belong to different electronic states, may differ considerably; thus a much stronger convergence to longer or shorter wavelengths may occur, leading to the characteristic head formation [when $\nu(m + 1) - \nu(m)$ from Eq. (84) vanishes]. The head is formed on the high-frequency side, in the R branch (the band is shaded to the red), when $B' < B''$, while the head is formed on the low-frequency side, in the P branch (the band is shaded to the

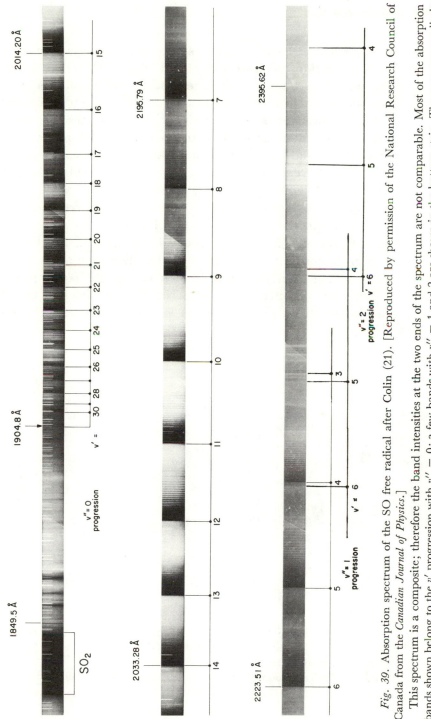

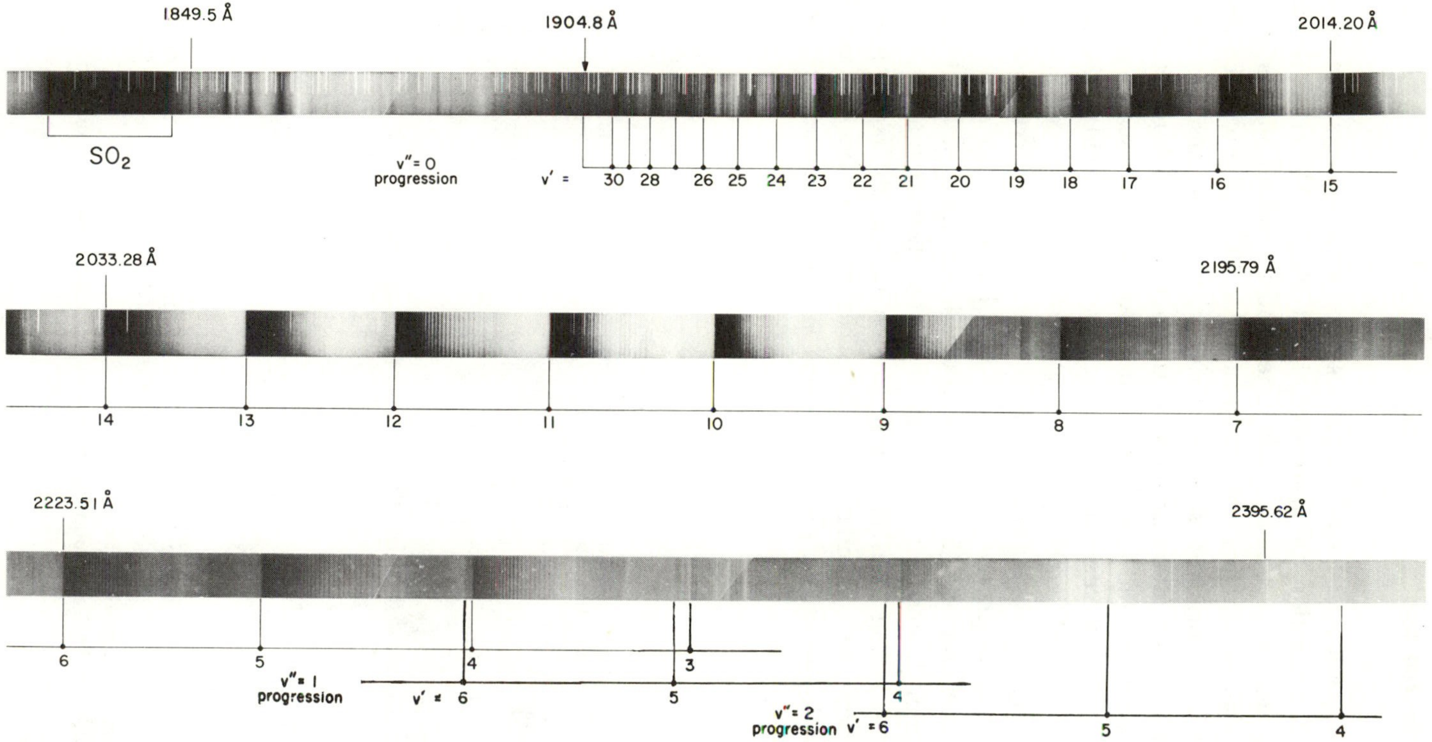

Fig. 39. Absorption spectrum of the SO free radical after Colin (21). [Reproduced by permission of the National Research Council of Canada from the *Canadian Journal of Physics*.]

This spectrum is a composite; therefore the band intensities at the two ends of the spectrum are not comparable. Most of the absorption bands shown belong to the v' progression with $v'' = 0$; a few bands with $v'' = 1$ and 2 are shown in the bottom strip. The convergence limit of the $v'' = 0$ progression is marked by a vertical arrow.

violet), when $B' > B''$. Figure 40 shows as an example of a $^1\Sigma$–$^1\Sigma$ transition an absorption band of the P_2 molecule as obtained by Creutzberg (24). The head formation and the zero gap (see p. 61) are clearly illustrated. In addition successive lines in the branches show a striking *alternation of intensities*. This intensity alternation arises from the fact that the P_2 molecule is homonuclear and has identical nuclei of spin $\frac{1}{2}$. In such a molecule, as we have seen earlier (p. 40), even and odd rotational levels have different symmetry with regard to an exchange of the nuclei, and, as a result of the nuclear spin, I, different statistical weights [in the ratio $I:(I+1)$]. Therefore, on account of the selection rule (55), even and odd lines in a Σ–Σ band have different intensities; that is, an intensity alternation arises.

The strong lines of the R branch of the P_2 band appear to form the continuation of the series of weak lines in the P branch [the strong lines have even m (odd J) in the R branch

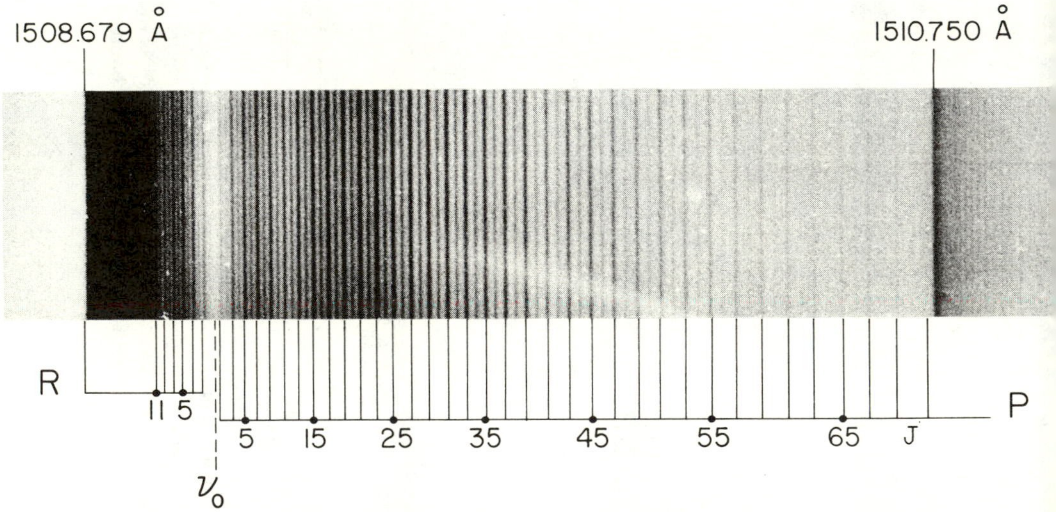

Fig. 40. Fine structure of a $^1\Sigma$–$^1\Sigma$ absorption band of the P_2 molecule after Creutzberg (24). [Reproduced by permission of the National Research Council of Canada from the *Canadian Journal of Physics.*]

Only the lines with odd J, which are the stronger ones, are marked.

but odd m (odd J) in the P branch]. For homonuclear molecules with $I = 0$ alternate lines would be missing. In a new spectrum of unknown origin it is clearly not obvious whether or not alternate lines are missing, but this can be established if one can ascertain whether or not the R lines form the continuation of the series of P lines (see below).

For $^2\Sigma - ^2\Sigma$ as well as $^3\Sigma - ^3\Sigma$ transitions the quantum number J has to be replaced by N. If the doublet or triplet splitting is very small and not resolved, the $^2\Sigma - ^2\Sigma$ and $^3\Sigma - ^3\Sigma$ transitions have the same rotational structure as $^1\Sigma - ^1\Sigma$. As an example, Fig. 41 (a) shows a spectrogram of the 0–0 band of the violet $^2\Sigma^+ - ^2\Sigma^+$ system of the CN radical as obtained in absorption at low temperature. The simple structure with a missing line at ν_0 is clearly seen. The doublet splitting is not resolved. In Fig. 41 (b) the same band is shown as obtained in emission at high temperature. Here lines

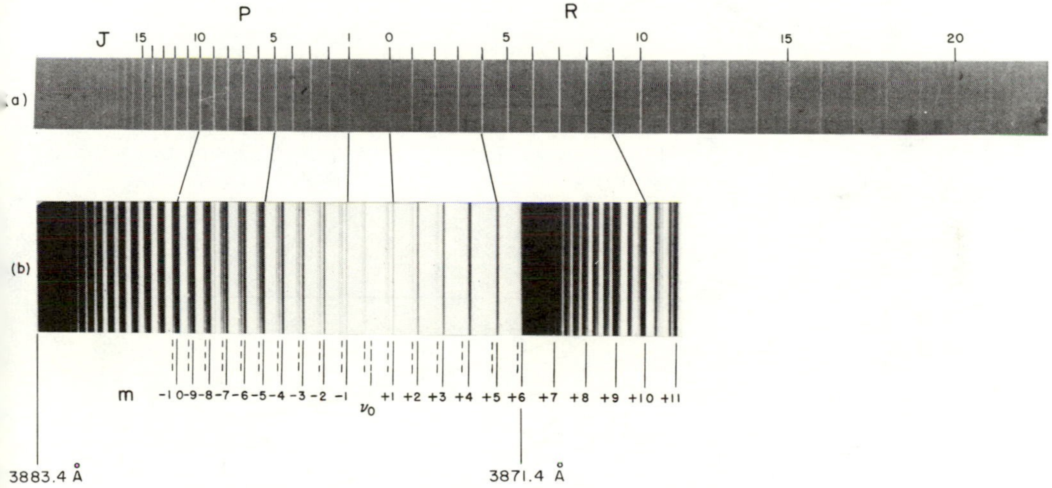

Fig. 41. Fine structure of the 0–0 band of the violet band system ($^2\Sigma - ^2\Sigma$) of the CN radical (a) in absorption at room temperature (b) in emission at higher temperature.

In (b) the lines of high J corresponding to the returning part of the P branch are visible and are resolved into doublets. The two spectra were taken with different spectrographs; therefore the scale is slightly different.

of high N in the returning part of the P branch are visible which clearly show a doubling in accordance with the facts that the electronic transition is ${}^2\Sigma$–${}^2\Sigma$ and that the doublet splitting in a ${}^2\Sigma$ state in a first approximation increases linearly with N [see Eq. (51a)].

The lines in a band are often represented in a *Fortrat diagram* in which J (or N) or m is plotted against ν. Such a diagram for the CN band of Fig. 41 is shown in Fig. 42. This figure shows how the band head arises in the P branch at $N = 28$ and how the P-lines of still higher N form a returning part of the branch, which interlaces with the lines of lower N of the P and R branches.

Another example of a Σ–Σ transition is shown in Fig. 43. This is a band of a new band system recently observed in both ab-

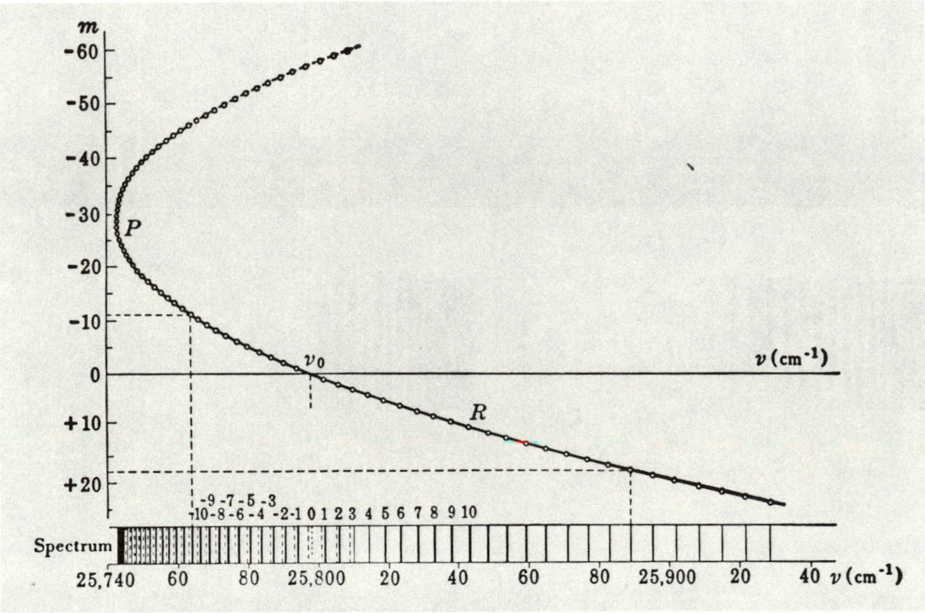

Fig. 42. Fortrat diagram of the CN band at 3883 Å (see Fig. 41) (from *MM* I, p. 48).

In addition to the Fortrat parabola a schematic spectrum is shown below to the same scale. The band origin is marked by a dotted line. No line is observed in the spectrum at this place; see Fig. 41(a).

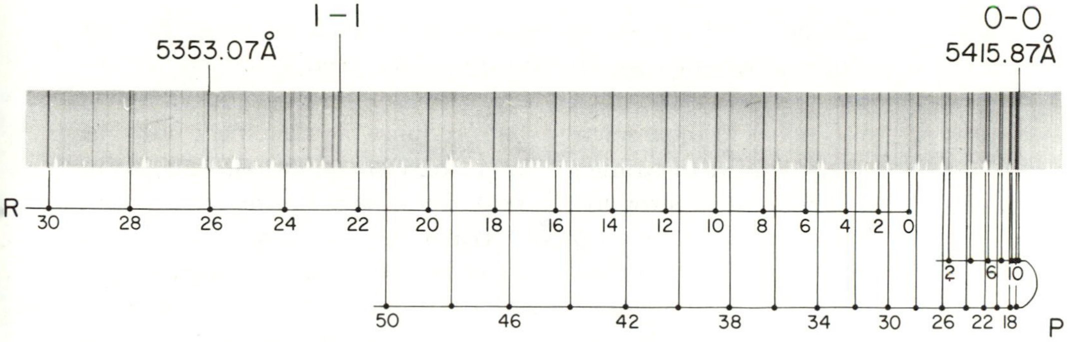

Fig. 43. Fine structure of the 0–0 band at 5416 Å of a new band system probably due to the C_2^- ion, in absorption; after Herzberg and Lagerqvist (66). [Reproduced by permission of the National Research Council of Canada from the *Canadian Journal of Physics.*]

The lines of the 1–1 band that overlaps the left-hand part of the spectrogram have not been marked.

sorption and emission in flash discharges through methane [Herzberg and Lagerqvist (66)]. This band again shows a single *P* and *R* branch. It can immediately be seen from the spectrogram that the lines of the *P* branch are not the continuation of those of the *R* branch but fall in the spaces between this continuation, showing conclusively that the spectrum is emitted by a molecule with two identical nuclei of spin zero. From the experimental conditions it is almost obvious that these nuclei must be carbon nuclei, and this was confirmed by an investigation with methane containing C^{13}. It therefore appeared likely that this spectrum was a new system of the C_2 radical. However, the vibrational and rotational analysis of the new system shows conclusively that neither the upper nor the lower state is a known state of C_2, and it is difficult to find among the predicted states of C_2 suitable ones to account for the two new states.

As a way out of this difficulty we proposed that the new spectrum is due to a C_2 ion, probably C_2^-. However, since C_2^- has an odd number of electrons (the same as CN), we should expect a doubling of the lines at high N values, just as in CN. No actual splitting has been observed, but a broadening at high N values is found which may be the beginning of such a doubling. Further confirmation of this identification has been obtained by the detection with a mass spectrometer of a fairly high concentration of C_2^- ions in the discharge that produces the new spectrum. The recent studies of Milligan and Jacox (96) on carbon radicals in a solid matrix strongly support the interpretation given here, but a fully conclusive proof has not yet been obtained. I have dwelt on this example in some detail, since it illustrates the methods and difficulties of identification of free-radical spectra.

Several examples of $^3\Sigma$–$^3\Sigma$ transitions of free radicals have been observed: those of SO (see Fig. 39), S_2, B_2, and Si_2. In all these radicals the ground state is $^3\Sigma^-$, and so the observed Σ–Σ transition must be $^3\Sigma^-$–$^3\Sigma^-$. The triplet splitting has been resolved in SO, S_2 and Si_2 but so far is unresolved in B_2. In this case the triplet nature of the transition is inferred from other evidence [Douglas and Herzberg (34)].

A $^4\Sigma^-$–$^4\Sigma^-$ transition is expected to occur for CH and NH^+ but has not yet been identified in either case, although for NH^+ a new spectrum, which might possibly be this predicted spectrum, has been observed by Narasimham and his colleagues (102).

Still another type of Σ–Σ transition has recently been studied for a number of free radicals: $^1\Sigma^+$–$^3\Sigma^-$. Such a transition appears to violate both the spin rule $\Delta S = 0$ and the rule $\Sigma^+ \not\leftrightarrow \Sigma^-$. However, the latter rule is valid only for transitions with $\Delta S = 0$, and therefore a $^1\Sigma^+$–$^3\Sigma^-$ transition, while indeed forbidden, is not forbidden so strongly that it would not be observed. In Fig. 44 the possible rotational transitions in such an electronic transition are indicated. It is seen that there are three branches with $\Delta N = 0$ called Q-form branches and one branch each with $\Delta N = +2$ and -2, called S- and O-form branches, respectively;[7]

[7] According to the standard notation true O, P, Q, R, S branches correspond to transitions with $\Delta J = -2, -1, 0, +1, +2$ respectively. Since the dependence of the rotational energy on N is almost the same as that on J (because by assumption the multiplet splitting is small) branches with $\Delta N = -2, -1, 0, +1, +2$ have the *form* of O, P, Q, R, S branches even when ΔJ is different from ΔN.

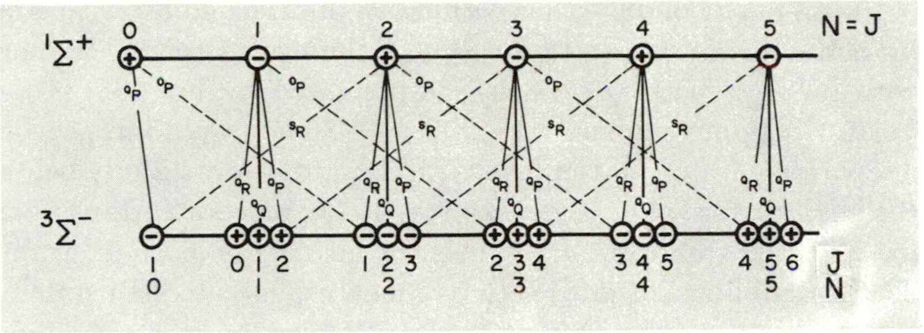

Fig. 44. Rotational transitions in a $^1\Sigma^+$–$^3\Sigma^-$ band.

The Q-form branches ($\Delta N = 0$) are indicated by solid lines, the S- and O-form branches ($\Delta N = \pm 2$) by broken lines. All branches are in accordance with the rules $\Delta J = 0, \pm 1$, and $+ \leftrightarrow -$; that is they are all P, Q, or R branches. The form of the branches determined by ΔN is indicated by the superscript preceding the main branch symbol.

in the latter the spacing of successive lines is approximately $4B$ instead of $2B$. An example of such a band structure is provided by the NF band in Fig. 45. This spectrum was obtained by Douglas and Jones (35) by the afterglow method described earlier. Similar spectra of the free radicals NBr, NCl, and SO have been observed by Milton, Dunford, and Douglas (97), Colin and Jones (22), and Colin (20).

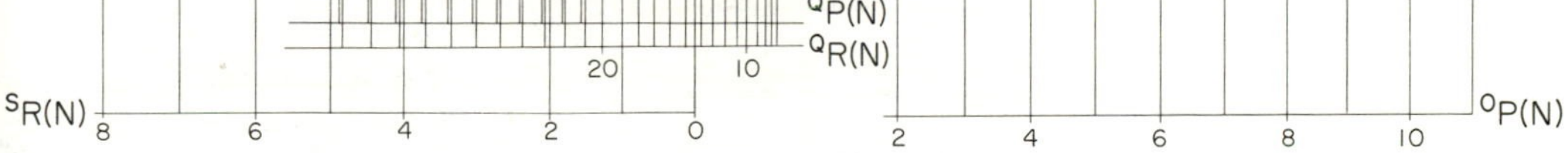

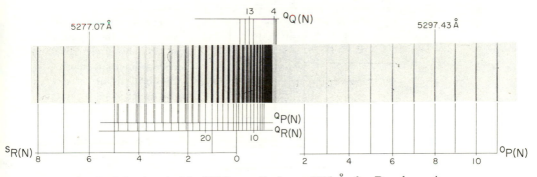

Fig. 45. Emission band of the NF free radical near 5290 Å after Douglas and Jones (35).

The widely spaced S- and O-form branches are clearly visible. Of the three Q-form branches ^QQ is rather weak.

In the bands of $^1\Pi$–$^1\Sigma$ transitions, in addition to the P and R branches corresponding to $\Delta J = \pm 1$, there is a strong Q branch with $\Delta J = 0$. This Q branch is represented by Eq. (85). In a Fortrat diagram similar to Fig. 42 a Q branch would appear as a separate parabola through ν_0 with its vertex only slightly below the abscissa axis, at $J = -\frac{1}{2}$. In Fig. 46 the rotational transitions for a $^1\Pi$–$^1\Sigma$ transition are shown in an energy-level diagram. The Λ doubling in the $^1\Pi$ state is much exaggerated. In spite of this doubling, on the basis of the parity selection rule (54), only three branches, P, Q, and R, arise. However, unless the Λ-type doubling is negligible, it causes a combination defect ϵ when the combination difference $R(J) - Q(J)$ is compared with $Q(J + 1) - P(J + 1)$; that is

$$R(J) - Q(J) = Q(J + 1) - P(J + 1) + \epsilon. \quad (102)$$

As Fig. 46 shows, the defect ϵ would be zero if there were no Λ doubling. From ϵ the magnitude of the Λ doubling can be

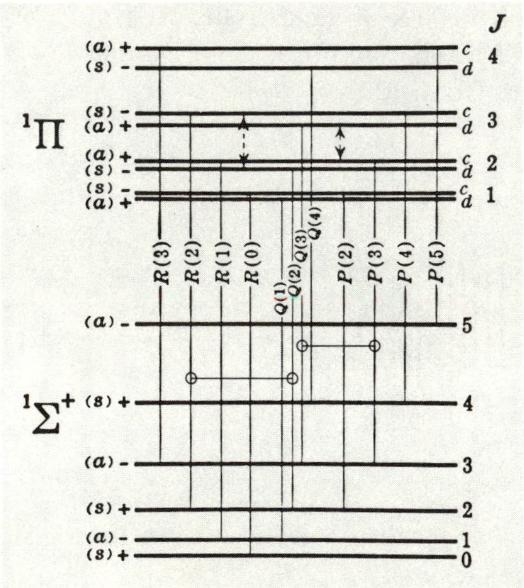

Fig. 46. Energy-level diagram for a $^1\Pi$–$^1\Sigma^+$ transition.

The wave-number differences of the transitions marked by circles give the energy intervals marked by the broken-line arrows in the upper state. They differ by the sum of the Λ doubling for $J = 2$ and 3 in the upper state [see Eq. (102)]. The designations (s) and (a) hold for a $^1\Pi_u$–$^1\Sigma_g^+$ transition of a homonuclear molecule.

determined. It is easily seen from Fig. 46 that here the first line of the P branch is not $P(1)$ as in a $^1\Sigma$–$^1\Sigma$ transition but $P(2)$, since the upper $^1\Pi$ state starts only with the $J = 1$ level. For the same reason the first line of the Q branch is $Q(1)$, while for the R branch, as before, $R(0)$ is the first line. An example of a $^1\Pi$–$^1\Sigma$ transition is the band of the BH free radical shown in Fig. 47; the three branches are clearly seen.

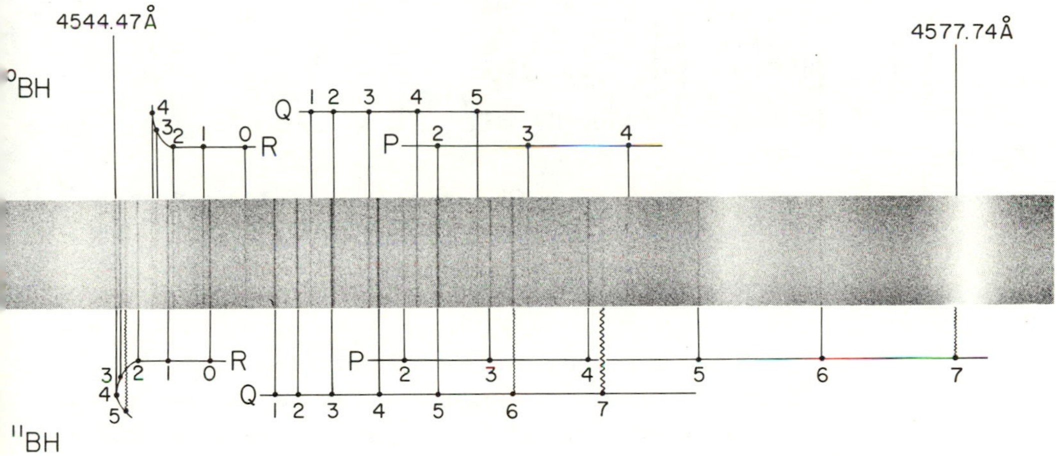

Fig. 47. The 3–3 band of the $A^1\Pi$–$X^1\Sigma^+$ transition of the BH radical after Johns, Grimm, and Porter (80). [Copyright 1967 by Academic Press, Inc., New York.]

Both the band of ^{11}BH (marked below) and of ^{10}BH (marked above the spectrogram) are visible. The last lines of the branches are slightly broadened by predissociation (see p. 196f.).

Bands of $^2\Pi$–$^2\Sigma$ transitions are more complicated than those of $^1\Pi$–$^1\Sigma$ transitions except when the $^2\Pi$ state belongs to Hund's case (b). In that case there is simply a doubling of the three branches. However, if the $^2\Pi$ state belongs to Hund's case (a), two subbands arise corresponding to $^2\Pi_{1/2}$–$^2\Sigma$ and $^2\Pi_{3/2}$–$^2\Sigma$. Each of these subbands will have six branches if the doublet splitting in the $^2\Sigma$ state is resolved. An example of such a transition is the red CN band shown in Fig. 48. A well-known example of a $^3\Pi$–$^3\Sigma$ transition with three subbands is the near-ultraviolet band of the NH free radical.

Σ–Π transitions are very similar to Π–Σ transitions. As a consequence of the exchange of upper and lower states

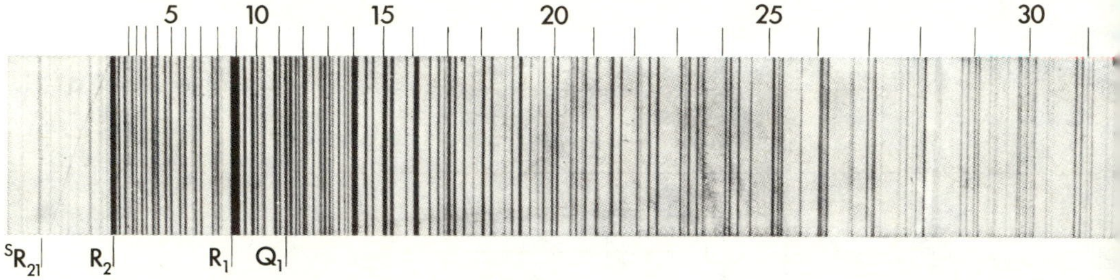

Fig. 48. Fine structure of the CN band at 5473 Å (9–3 band of the "red" $^2\Pi$–$^2\Sigma$ system) after Jenkins, Roots, and Mulliken (77) (from *MM* I, p. 47). Only the lines of the R_2 branch are indicated above the spectrogram. Below, the four heads are marked.

for $^1\Sigma$–$^1\Pi$ bands the line $R(0)$ is missing rather than $P(1)$ in $^1\Pi$–$^1\Sigma$ bands. Many Σ–Π transitions of free radicals are known. As an example we show in Fig. 49 a near-ultraviolet absorption band of CH belonging to a $^2\Sigma^-$–$^2\Pi$ transition. It shows only six branches, since the $^2\Pi$ state belongs to Hund's case (b).

In Π–Δ and Δ–Π transitions the band structure is similar to the corresponding Σ–Π and Π–Σ transitions, except that one more line is missing at the beginning of all branches and that the Λ doubling, since it occurs in both the upper and the lower state, now produces a doubling in all branches, in addition to any spin doubling or tripling that may be present. The familiar CH band at 4315 Å which occurs in every Bunsen flame is an example of a $^2\Delta$–$^2\Pi$ transition. For Π–Π and Δ–Δ transitions the band structure is similar to Π–Δ transitions except that the Q branch is weak and its intensity decreases rapidly with increasing J. The Swan bands of the C_2 radical in the visible region which occur in every carbon arc (see Fig. 1) are an example of a $^3\Pi$–$^3\Pi$ transition.

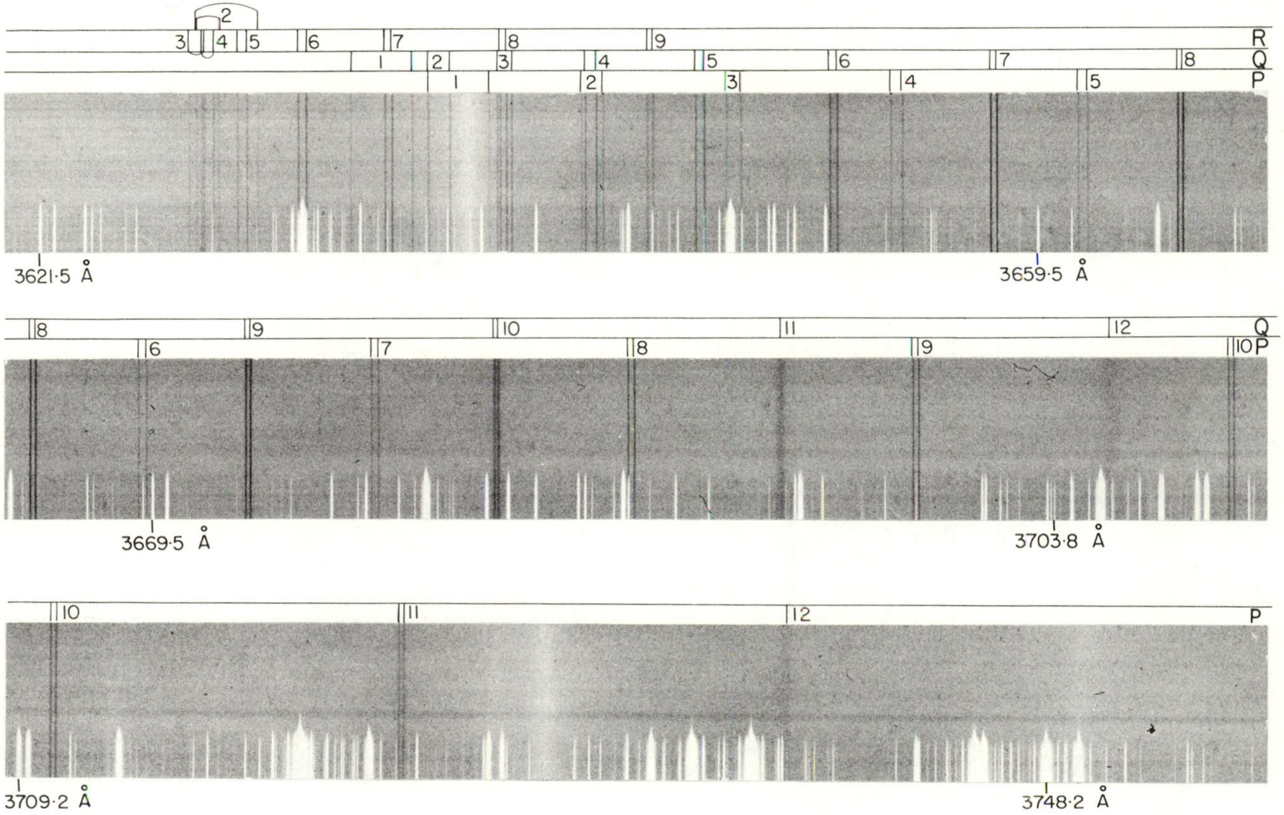

Fig. 49. Fine structure of the CH band at 3630 A (1–0 band of the $B^2\Sigma^- - X^2\Pi$ system) as obtained in absorption by Herzberg and Johns (65). [Reproduced by permission of the University of Chicago Press from the *Astrophysical Journal*.]

The doublet structure of the lines is clearly visible. The lines of highest J are broadened by predissociation (see p. 196).

The rotational and vibrational analysis of the observable band systems of a given free radical allows us to establish various

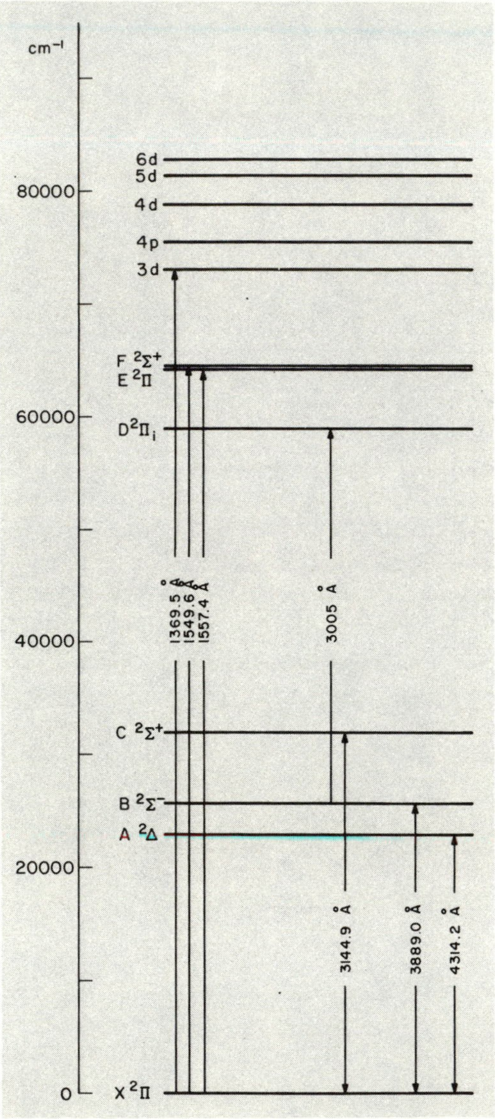

Fig. 50. Energy-level diagram of the observed electronic states of the CH free radical.

The observed transitions are shown as vertical arrows.

electronic states of this radical. When Rydberg series of such states can be found, an ionization potential of the radical can be derived. This has been done for the CH radical, for example, for which in Table 2 (p. 33) the Rydberg states predicted from molecular-orbital theory are listed. In Fig. 50 the observed electronic states, which include a Rydberg series, are shown in an energy-level diagram. The ionization potential derived from the Rydberg series is 10.64 eV. Table 8 lists the molecular constants of the various electronic states of CH as derived from the study of its band spectrum [Herzberg and Johns (65)]. As a

Table 8. Molecular Constants of CH in its Various Electronic States

State	T_0	$\Delta G(\frac{1}{2})$	B_0	r_0
na	82788			
6*d*	82726			
na	81545			
5*d*	81271			
na	81006			
4*d*	78660			
4*p*	75550			
$3d(^2\Sigma, {}^2\Pi, {}^2\Delta)$	72960			
$F\ ^2\Sigma^+$	64531.5		12.17	1.221
$E\ ^2\Pi$	64211.7		12.6	1.2_0
$D\ ^2\Pi_i$	58981.0		13.7	1.1_2
$C\ ^2\Sigma^+$	31778.1	2612.5	14.2466	
$B\ ^2\Sigma^-$	25698.2	1794.9	12.645	
$A\ ^2\Delta$	23217.5	2737.4	14.577	
$X\ ^2\Pi_r$	0	2732.50	14.190	

The states marked na are not assigned.

second example, Fig. 51 shows an energy-level diagram of all the observed electronic states of BH (7)(80).

Another way of presenting the data obtained from the spectrum for a given free radical is to derive from the observed molecular constants the potential functions of each observed state and plot them in a diagram. This was done for the C_2 free radical by Ballik and Ramsay (4), whose diagram we reproduce in Fig. 52. More recently three additional electronic states have been found [Herzberg, Lagerqvist, and Malmberg (67)], which are not included in the diagram.

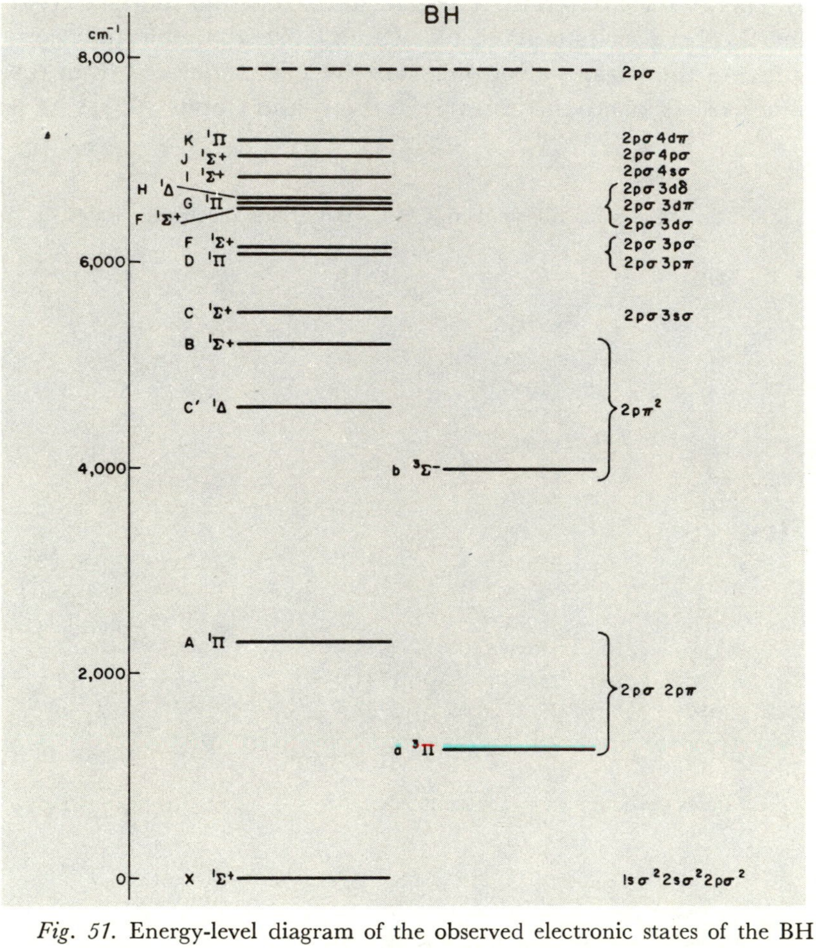

Fig. 51. Energy-level diagram of the observed electronic states of the BH free radical.

The relative positions of singlet and triplet states are not yet known. At the right the electron configurations are given.

Nothing has been said in the preceding discussion about various methods to determine the best values of the rotational and vibrational constants. For these topics we refer to MMI p. 175ff.

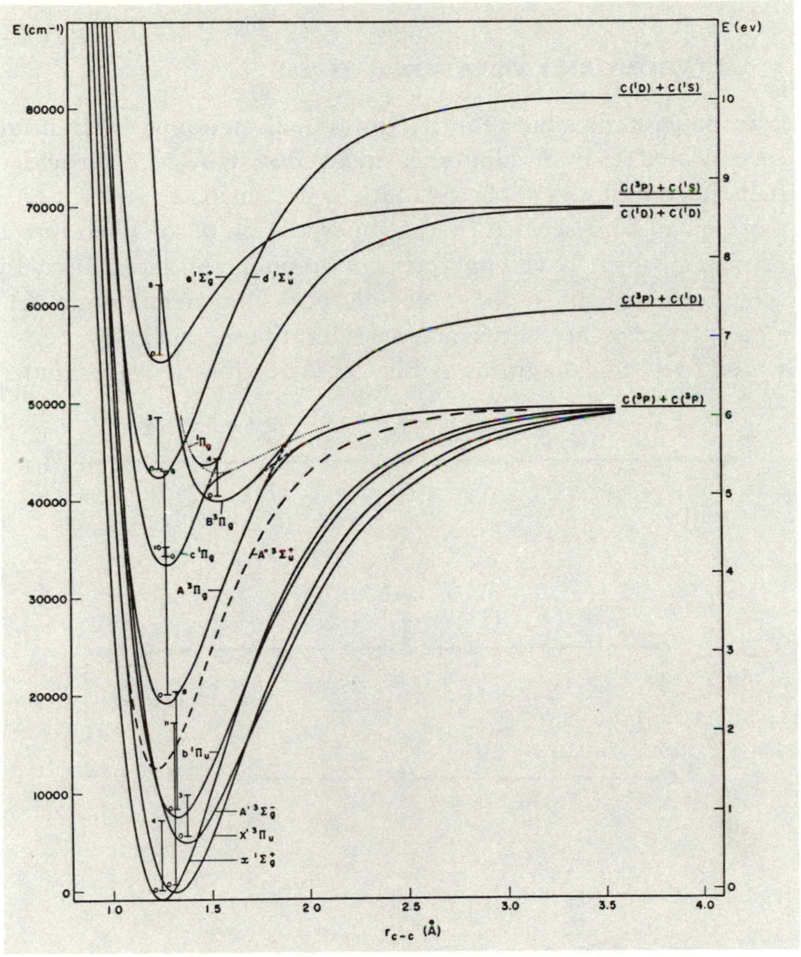

Fig. 52. Potential functions of the observed electronic states of the C_2 free radical after Ballik and Ramsay (4). [Reproduced by permission of the University of Chicago Press from the *Astrophysical Journal.*]

Three recently discovered electronic states [Herzberg, Lagerqvist, and Malmberg (67)] are not included.

III. LINEAR POLYATOMIC RADICALS AND IONS

A. VIBRATIONS AND VIBRATIONAL LEVELS

In polyatomic molecules the vibrational motion is much more complicated than in diatomic molecules. However, provided that the amplitudes of the oscillations are small, any vibrational motion can be described by the superposition of so-called *normal vibrations*, which have characteristic frequencies determined by the force constants in the molecule. For a linear triatomic XY_2 molecule there are three normal vibrations, which are represented by vector diagrams in Fig. 53. The vibrations designated

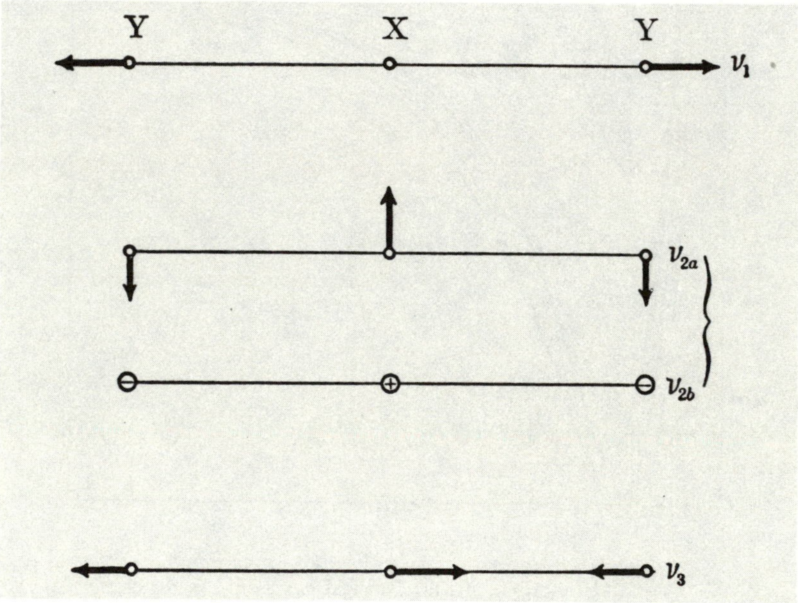

Fig. 53. Normal vibrations of a linear XY_2 molecule (schematic) (from *MM* II, p. 66).

The displacements of the nuclei during the vibrations are indicated by the heavy arrows (not to scale).

ν_1 and ν_3 are stretching vibrations along the line of the internuclear axis, while ν_2 is a bending vibration in which the nuclei oscillate perpendicular to the internuclear axis. This vibration has two degrees of freedom; it is doubly degenerate because it may take place in the plane of Fig. 53 or in a plane perpendicular to it or, for that matter, in any other plane through the internuclear axis. All these vibrations in various planes can be represented as a superposition of vibrations in two perpendicular planes.

The vibrations ν_1 and ν_3 in a linear symmetric XY_2 molecule have different symmetry. The first is totally symmetric—that is, it remains unchanged for any of the symmetry operations that leave the molecule unchanged; the symmetry type (species) of ν_1 is represented by Σ_g^+. The stretching vibration ν_3 is antisymmetric with respect to the center of symmetry, and its symmetry type is represented by Σ_u^+. The bending vibration ν_2 is doubly degenerate and has the symmetry type Π_u. In analogy to the usage for electronic orbitals and electronic states we use small letters for individual normal vibrations [for example, $\nu_1(\sigma_g^+)$, $\nu_2(\pi_u)$, $\nu_3(\sigma_u^+)$] and capital letters for the resulting vibrational state (for example, when both ν_1 and ν_3 are singly excited the vibrational state is a Σ_u^+ state).

In an unsymmetric linear triatomic molecule of the form XYZ there are also three normal vibrations, but the two stretching vibrations ν_1 and ν_3 are no longer distinguished by symmetry, since there is no longer a center of symmetry; they are both of species Σ^+. For a four-atomic linear molecule there would be three stretching vibrations and two bending vibrations (see *MM* II, p. 181). More generally, for an N-atomic linear molecule there would be $N - 1$ stretching vibrations and $N - 2$ bending vibrations, all of the latter being doubly degenerate.

The vibrational energy levels are in a first approximation like those of a group of independent harmonic oscillators; that is, the vibrational term value is given by

$$G(v_1, v_2, v_3, \ldots) = \sum_i \omega_i \left(v_i + \frac{d_i}{2}\right), \tag{103}$$

where ω_i is the vibrational frequency (in cm^{-1}) of the normal vibration ν_i, where v_i is the corresponding vibrational quantum number, and d_i is 1 or 2 depending on whether the vibration is nondegenerate or doubly degenerate.

If we want to take account of the interaction of the vibrations and of their anharmonicity, we must go to the next order of approximation. Then we obtain for the *vibrational term-value* [corresponding to Eq. (18) for diatomic molecules]

$$G(v_1, v_2, v_3, \ldots) = \sum_i \omega_i\left(v_i + \frac{d_i}{2}\right) + \sum_i \sum_{k \geq i} x_{ik}\left(v_i + \frac{d_i}{2}\right)\left(v_k + \frac{d_k}{2}\right) + \cdots + \sum_i \sum_{k \geq i} g_{ik} l_i l_k, \tag{104}$$

in which the x_{ik} and g_{ik} are anharmonicity constants. The last term in Eq. (104) applies only to degenerate vibrations and arises because these degenerate vibrations have a (vibrational) angular momentum given by $l_i h/2\pi$, where

$$l_i = v_i,\ v_i - 2, \ldots, 1 \text{ or } 0. \tag{105}$$

That an angular momentum can arise for a degenerate vibration is easily seen classically if the two component vibrations of ν_2 in Fig. 53 are simultaneously excited with equal amplitude but with a phase difference of 90°. As a result each of the nuclei carries out a rotational motion about the internuclear axis in planes perpendicular to the axis; all nuclei move in the same sense—that is, an angular momentum about the internuclear axis arises. For all nondegenerate vibrations the angular momentum $\boldsymbol{l}_i$ is zero, and therefore when degenerate vibrations are not excited ($v_i = 0$, $l_i = 0$), the last term in Eq. (104) can be omitted.

It should be mentioned that Eq. (104) does not give a complete representation of the vibrational levels when two levels of the same symmetry type lie close together. In that case *perturbations*—small shifts of these levels—arise that are not represented by Eq. (104). The first example of such a perturbation was recognized by Fermi in CO_2 (where $\nu_1 \approx 2\nu_2$), and these perturbations are usually referred to as *Fermi resonances.*

The angular momentum $\boldsymbol{l}$ of a degenerate vibration is an angular momentum about the axis of symmetry just like the orbital angular momentum $\boldsymbol{\lambda}_i$ of an electron. This analogy is useful for an understanding of the symmetry types (species) of the vibrational wave functions of the excited levels of a bending vibration. These symmetry types are indicated in Fig. 54 for $v_2 = 0, 1, 2, 3, 4$. The first vibrational level, $v_2 = 1$, has the same species (Π) as the normal vibration—that is, there is a vibrational angular momentum of one unit; in this state we can never have

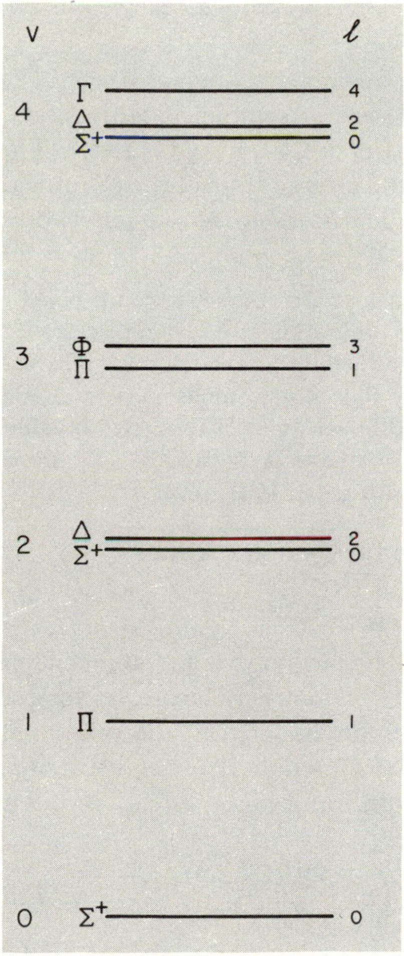

Fig. 54. Splitting of the higher vibrational levels of a bending (π) vibration of a linear molecule.

For symmetric molecules (point group $D_{\infty h}$) the levels are alternately g and u for even and odd (or odd and even) v.

zero angular momentum. In the second vibrational state, $v_2 = 2$, we have according to Eq. (105) two sublevels with $l_2 = 2$ and 0 of species Δ and Σ^+. They correspond to parallel and antiparallel orientations of the two component vectors with $l_2 = 1$, just as two π electrons can have their $\lambda_i = 1$ only parallel or antiparallel. The two sublevels Δ and Σ^+ would coincide if the vibrations were strictly harmonic, but the last term in Eq. (104) (which is caused by anharmonicity) gives rise to a slight difference in energy. For the third vibrational state, $v_2 = 3$, the quantum number l_2 is 3 or 1, corresponding to a Φ and a Π vibrational sublevel, and so on.

For a molecule with a center of symmetry, such as linear XY_2, the vibrational levels for even v_2 are *g* (symmetric with respect to inversion) while those for odd v_2 are *u* (antisymmetric). This is the situation, for example, in the ground state of the C_3 radical, which is linear and for which levels up to $v_2 = 6$ have been observed.

When two or more bending vibrations are excited, which is possible only for four-atomic (or more than four-atomic) linear molecules, we must form the resultant $\boldsymbol{L}$ of the individual $\boldsymbol{l}_i$. We consider only one example. Suppose that in a linear four-atomic molecule $v_4 = 1$ and $v_5 = 1$; that is, $l_4 = 1$ and $l_5 = 1$. The resultant of these two l_i values can be either 2 or 0^+ or 0^-, corresponding to Δ, Σ^+, and Σ^- vibrational states, respectively. For further details see *MM* II, p. 211f.

B. ROTATIONAL LEVELS, INTERACTION OF ROTATION AND VIBRATION

(1) Nondegenerate vibrational levels

The rotational motion of a linear polyatomic molecule is essentially the same as that of a diatomic molecule, as long as the vibrational levels are nondegenerate; that is, the rotational-energy formula is given by Eq. (24), where now the subscript v corresponds to the *set* of quantum numbers $v_1, v_2, v_3, \ldots$. The rotational constant B_v is given by

$$B_v = B_e - \sum \alpha_i \left(v_i + \frac{d_i}{2}\right) + \cdots; \qquad (106)$$

where as previously $d_i = 1$ or 2 for nondegenerate and doubly degenerate vibrations respectively. There are as many rotational constants α_i as there are normal vibrations [compare Eq. (21)]. The equilibrium constant B_e is given by

$$B_e = \frac{h}{8\pi^2 c I_e}, \tag{107}$$

where I_e is the moment of inertia of the molecule in its equilibrium position. The rotational constant B_0 in the lowest vibrational level is

$$B_0 = B_e - \tfrac{1}{2} \sum d_i \alpha_i + \cdots \tag{106a}$$

The contributions of each degenerate vibration to $B_e - B_0$ is α_i while that of each nondegenerate vibration is $\frac{1}{2}\alpha_i$ as for diatomic molecules. This result arises because for the degenerate vibrations there is a whole quantum $h\nu_{\text{osc}}$ of zero-point vibration, for the nondegenerate ones only half a quantum.

The effect of the electron spin on the rotational levels (*spin-splitting*) is entirely similar to the diatomic case (see p. 42f.), and the same applies to the interaction of the orbital angular momentum Λ of the electrons with the rotation, which leads as before to *Λ-type doubling*.

(2) Degenerate vibrational levels

In a degenerate vibrational level of type Π, Δ, . . . we have to use as rotational energy formula the analogue of (44), in which Λ has been replaced by l_i [or, when several degenerate vibrations are excited, by $L = \Sigma\,(\pm\, l_i)$]—that is,

$$F_v(J) = B_v[J(J+1) - l_i^2]. \tag{108}$$

Again, very often the term $-B_v l_i^2$, since it is fixed for a given vibrational level, is included in the vibrational energy, and thus the same rotational formula can be used as for nondegenerate vibrational levels. However, one modification must often be considered, particularly for Π vibrational levels, namely, *l-type doubling* [the analogue of Λ-type doubling (see p. 42)], which is brought about by the interaction of the vibrational angular momentum $\boldsymbol{l}$ with the rotation of the molecule. In Π vibrational levels the magnitude of the l-type doubling, just like that of Λ-type doubling, is given by

$$\Delta\nu = q_i J(J + 1), \tag{109}$$

where the splitting constant q_i can be fairly simply evaluated from the rotational and vibrational constants of the electronic state considered.

For a symmetric linear triatomic molecule the splitting constant q_2 is given by (see *MM* III, p. 70)

$$q_2 = \frac{B_e^2}{\omega_2}\left(1 + \frac{4\omega_2^2}{\omega_3^2 - \omega_2^2}\right)(v_2 + 1). \tag{110}$$

According to this formula, if the rotational constant B_e and the zero-order frequencies ω_2 and ω_3 of the two normal vibrations ν_2 and ν_3 are known, the value of the splitting constant can be predicted, and in most cases agreement with observation has been found. There are two contributions to the splitting constant: the first term (B_e^2/ω_2) is a harmonic term, while the second term is caused by *Coriolis interaction* between the vibrations ν_2 and ν_3. The magnitude of q_2 is in general substantially larger than the splitting constant q for Λ-type doubling. Thus l-type doubling is a fairly important phenomenon in the study of linear polyatomic radicals.

If there are several vibrations with which the bending vibration considered can interact, as is the case for unsymmetric linear triatomic molecules and for four- and five-atomic linear molecules, then Eq. (110) must be replaced by

$$q_i = \frac{B_e^2}{\omega_i}\left(1 + 4\sum_k \frac{\zeta_{ik}^2\omega_i^2}{\omega_k^2 - \omega_i^2}\right)(v_i + 1), \tag{111}$$

where ζ_{ik} are Coriolis coefficients which depend in a fairly complicated way on the masses and potential constants, and where the summation is over all vibrations ν_k with which Coriolis interaction arises.

The l-type doubling in Δ vibrational states is extremely small except when a Σ vibrational state lies close by. As Fig. 54 shows, this is usually the case for $v_2 = 2$, as long as the constant g_{22} is not too large. In that case one of the components of the Δ state interacts with the Σ state and a splitting that is proportional to $J^2(J + 1)^2$ arises, while at the same time the Σ state has an anomalously large D_v value. This phenomenon, called *l-type resonance*, has been of some significance in the study of the ground state of the C_3 free radical [see Gausset et al. (43)].

C. INTERACTION OF VIBRATION AND ELECTRONIC MOTION (VIBRONIC INTERACTION)

(1) Nondegenerate electronic states

In nondegenerate electronic states the interaction of vibration and electronic motion is entirely similar to that in diatomic molecules. It is taken into account by the fact that the ω_i and x_{ik} in Eq. (104) correspond to the potential function of the electronic state considered—that is, by the fact that the electronic energy determines the potential in which the nuclei move. This statement implies the assumption that the Born-Oppenheimer approximation is valid. The vibrational-electronic energy or, for short, the *vibronic* energy, is in this case to a good approximation simply the sum of electronic and vibrational energy:

$$E_{ev} = E_e + E_v, \tag{112}$$

or, in term values,

$$T_{ev} = T_e + G(v_1, v_2, \ldots). \tag{113}$$

The corresponding vibronic eigenfunctions are

$$\psi_{ev} = \psi_e(q, 0)\psi_v(Q), \tag{114}$$

where q stands for all the electronic, Q for all the nuclear coordinates. In general the electronic eigenfunction $\psi_e(q, Q)$ depends on the nuclear coordinates as parameters, but in the Born-Oppenheimer approximation we use the eigenfunction $\psi_e(q, 0)$ for the equilibrium position ($Q = 0$). The symmetry types or species of the vibronic eigenfunctions ψ_{ev} are simply the "products" of the electronic and the vibrational species. For example, if the vibrational species is Σ_u^+ and the electronic species is Σ_g^-, then the vibronic species is Σ_u^-; if the vibrational species is Π_u again for the electronic species Σ_g^-, then the vibronic species is also Π_u.

(2) Degenerate electronic states: singlets

In a degenerate electronic state, in general, for a given vibrational level several vibronic species arise; that is, there are several vibronic levels for a given combination of electronic and vibrational states. In Fig. 55 these sublevels are shown for the various vibrational levels of the vibration ν_2 of a linear triatomic molecule in a Π_g electronic state. The electronic and the vibrational angular momentum form a resultant $\boldsymbol{K}$, which may be called the *vibronic angular momentum* (exclusive of electron spin). The corresponding quantum number (which determines the vibronic species) is given by

$$K = |\pm\Lambda \pm l|. \tag{115}$$

From this simple formula the results shown in Fig. 55 are easily derived. Here it must be noted that the value $K = 0$ always arises in two ways, and the two corresponding eigenfunctions

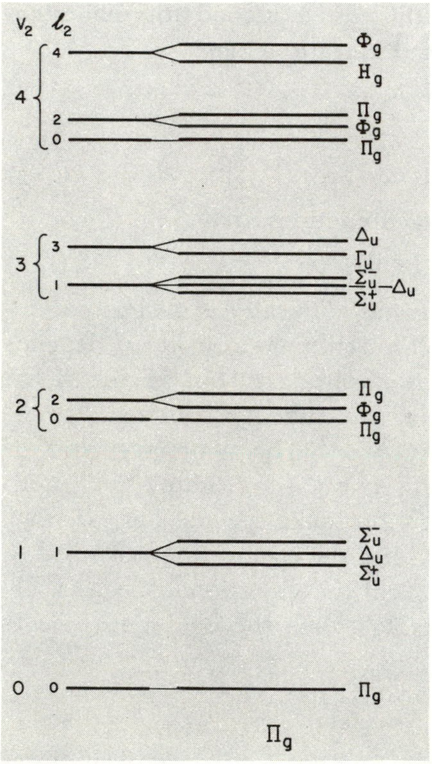

Fig. 55. Vibronic splittings in a Π_g electronic state of an XY_2 molecule upon excitation of the bending vibration ν_2.

At the left the energy levels are given for zero vibronic interaction, as in Fig. 54. The anharmonicity splittings are greatly exaggerated. The vibronic splittings at the right are not plotted to scale. They are often larger than shown.

are the sum and difference of the "original" ones, giving rise to Σ^+ and Σ^- vibronic levels.

If we include in the wave equation those terms that are neglected when the product (114) is assumed to be the solution —that is, if we take account of vibronic interaction—levels of given v_2, l_2 in Fig. 55 that are coincident without this interaction (as shown at the left), will in fact be split, and we obtain the manifold of vibronic levels as shown at the right in Fig. 55.

For a quantitative evaluation of the vibronic splittings we must consider the variation of the potential energy with the bending coordinate. As was first recognized by Teller (69) and worked out in detail by his student Renner (120), the potential function in a degenerate electronic state splits into two when the molecule is bent, as shown in Fig. 56(a). This splitting is the

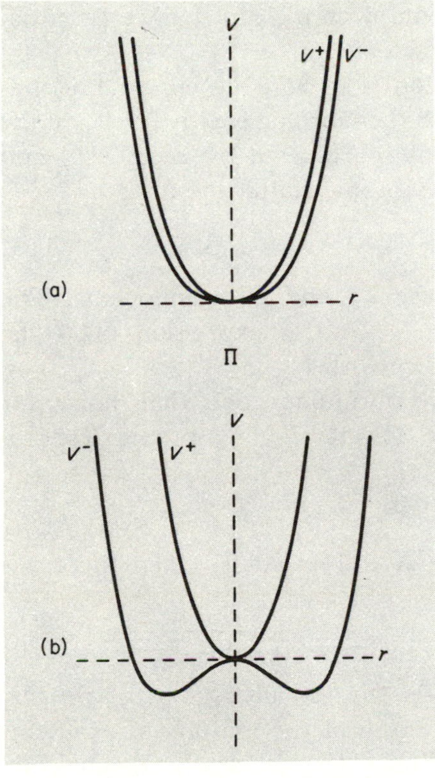

Fig. 56. Potential functions for the bending vibration in Π electronic states of linear molecules (a) for small and (b) for large vibronic interaction (from *MM* III, p. 27).

The abscissa is the bending coordinate.

basis of what we shall call here the Renner-Teller effect. In zero-order approximation—that is, without taking account of vibronic interactions—the potential function, which for reasons of symmetry is an even function of the bending displacement r, can be represented by the expression

$$V^0 = ar^2 + br^4 + \cdots, \tag{116}$$

where a and b are potential constants. The splitting into V^+ and V^- produced by the vibronic interaction can be represented by a similar equation

$$V^+ - V^- = \alpha r^2 + \beta r^4 + \cdots, \tag{117}$$

where in general α and β are small compared to a and b. However, if the interaction is so large that $\frac{1}{2}\alpha > a$, then the two potential functions will look like those shown in Fig. 56(b); the lower one corresponds to a nonlinear molecule, since potential minima arise only for nonzero r.

For small vibronic interaction [Fig. 56(a)] Renner (120) has developed explicit formulae for the vibronic energy levels, on the basis of the potential functions represented by Eqs. (116) and (117). For $K = 0$—that is, Σ vibronic states—he finds

$$G^{\pm}(v_2, 0) = \omega_2\sqrt{1 \pm \epsilon}\,(v_2 + 1), \qquad v_2 = 1, 3, 5, \ldots, \tag{118}$$

where the $\pm$ sign refers to the Σ^+ and Σ^- components. The vibrational quantum number v_2 in this expression takes the values 1, 3, 5, . . . in accordance with Fig. 55.

For $K \neq 0$, there are two formulae: one that holds for $v_2 = K - 1$—that is, for the lowest single vibronic level of species Π, Δ, . . . (see Fig. 55):

$$G(v_2, K) = \omega_2[(v_2 + 1) - \tfrac{1}{8}\epsilon^2 K(K + 1)], \qquad v_2 = K - 1; \tag{119}$$

the other that holds for $v_2 > K - 1$—that is, when there are two levels for each v_2:

$$G^{\pm}(v_2, K) = \omega_2(1 - \tfrac{1}{8}\epsilon^2)(v_2 + 1) \pm \tfrac{1}{2}\omega_2\epsilon\sqrt{(v_2 + 1)^2 - K^2}. \tag{120}$$

In these formulae $\epsilon = \alpha/2a$ is the so-called *Renner parameter*, which is a measure of the strength of the vibronic interaction. The quantities $G(v_2, K)$ from Eqs. (118, 119, and 120) should be substituted for

$$\omega_2(v_2 + 1) + g_{22}l_2^{\,2}$$

in the previous equation (104) for the vibrational levels of a nondegenerate electronic state. In Fig. 57 the vibronic levels in a Π electronic state as calculated from the above formulae are represented to scale in an energy-level diagram. The vibronic splittings of levels with given v_2 and l are also called *Renner-Teller splittings*.

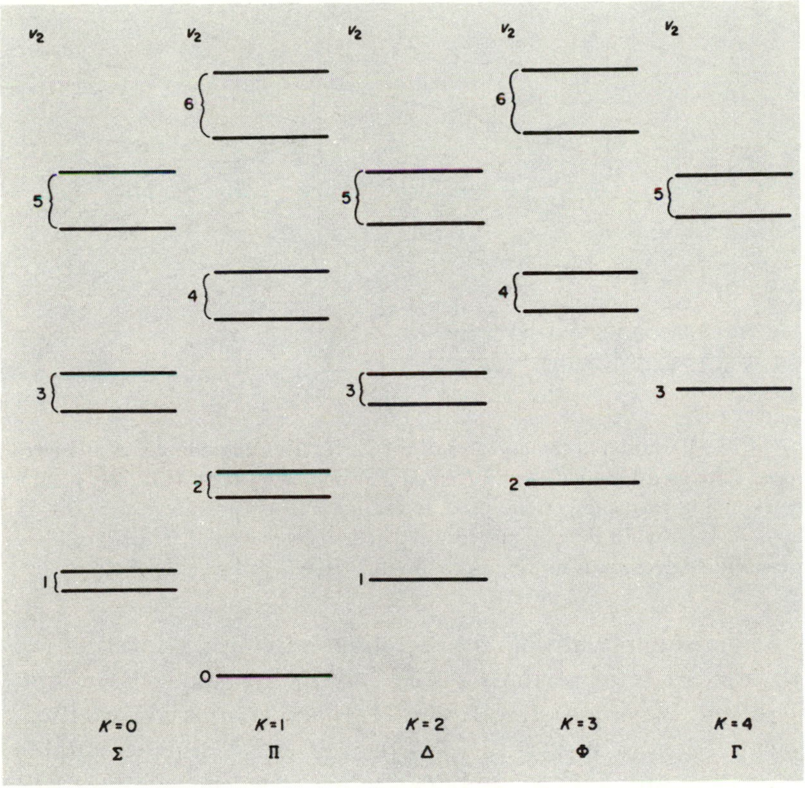

Fig. 57. Vibronic levels in a Π electronic state of a linear molecule for $\epsilon = 0.1$ (from *MM* III, p. 32).

All levels except those with $K = v_2 + 1$ occur in pairs. They are plotted to scale neglecting anharmonicity.

With increasing ϵ, a crossing between Σ^+ and Σ^- vibronic levels of different v_2 arises, as shown in Fig. 58 at the left. Similarly, intersections arise between the two Π and between the two Δ states that arise from Eq. (120). However, while the intersections of Σ^+ and Σ^- are required by symmetry, for the Π and Δ states ($K = 1$, $K = 2$) the intersections are actually avoided, as shown in Fig. 58 center and right. The upper set of levels belongs to the upper potential function in Fig. 56, the lower set to the lower potential function, but this correlation is exact only for Σ^+ and Σ^-. For the Π, Δ, . . . vibronic levels it is a good approximation only for small ϵ.

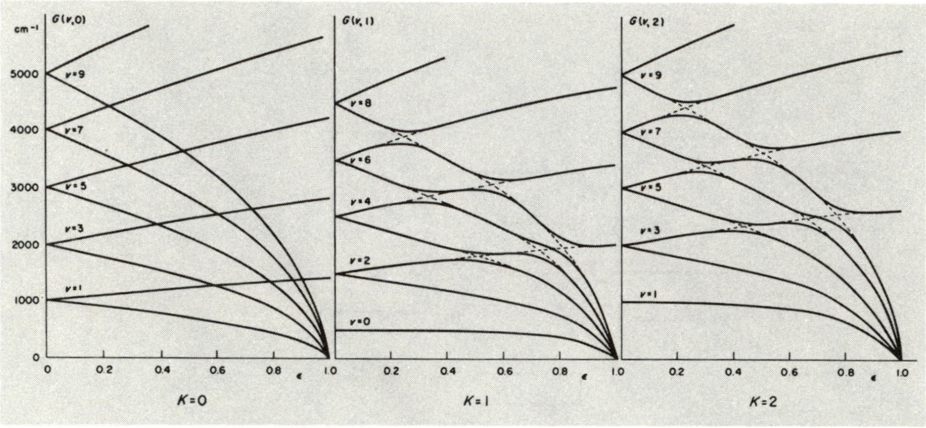

Fig. 58. Variation of the energies of the Σ, Π, Δ vibronic levels of a Π electronic state as a function of the Renner parameter ϵ (from *MM* III, p. 31). A bending frequency of 500 cm^{-1} is assumed. The lowest level $v = 0$ is in the $K = 1$ set (it is a Π level) and for $\epsilon = 0$ occurs at 500 cm^{-1}. The broken-line correlation corresponds to the approximation in which Eq. (120) holds.

As an example, in Fig. 59 the observed vibronic levels of the $^1\Pi_u$ excited state of the C_3 free radical are shown [Gausset, Herzberg, Lagerqvist, and Rosen (43)] and compared with those calculated from Eqs. (118), (119), and (120) using a suitable value of ϵ. Although a number of vibronic levels have not been observed, those that have been observed do fit the calculations fairly well. We shall not consider here the formulae that apply when vibronic interaction is large (see *MM* III, pp. 30, 32).

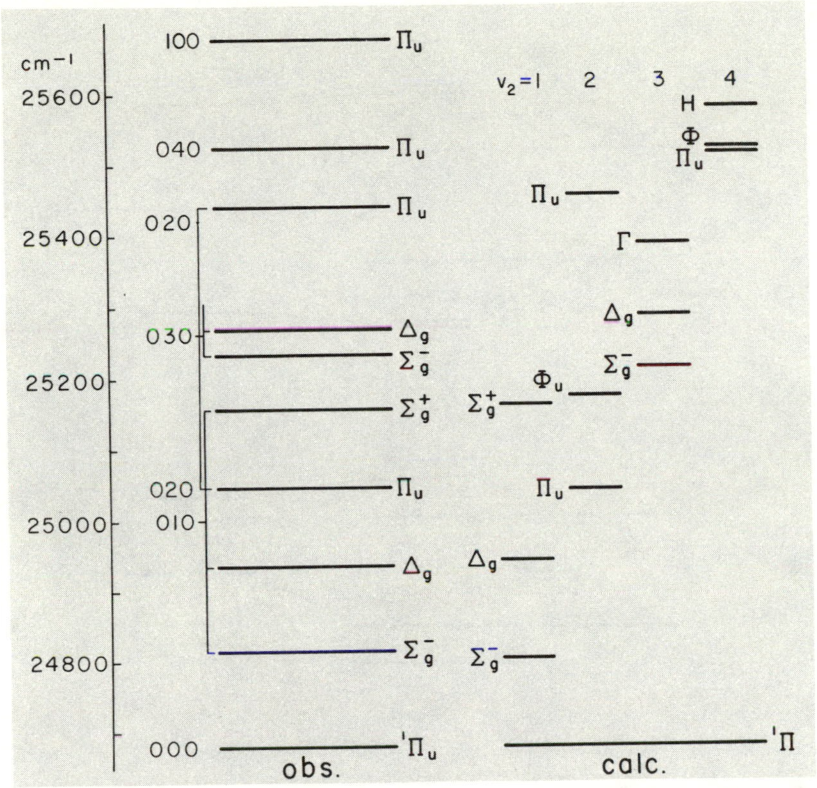

Fig. 59. Observed and calculated vibronic levels in the $^1\Pi_u$ excited state of the C_3 free radical.

Some of the predicted levels for $v_2 = 3$ and 4 lie beyond the range of the figure; they have not been observed.

(3) Degenerate electronic states: doublets

In multiplet Π and Δ states, the effect of vibronic interaction is more complicated. We shall only consider briefly the case of $^2\Pi$. For small spin-orbit coupling [Hund's case (b)] everything is essentially the same as for singlets. This is shown at the right in Fig. 60 in the column marked $A = 0$. At the left in the same figure in the column marked $\epsilon = 0$ are given the vibronic levels produced by spin-orbit coupling alone, which is assumed to be fairly large. When both spin-orbit coupling and vibronic inter-

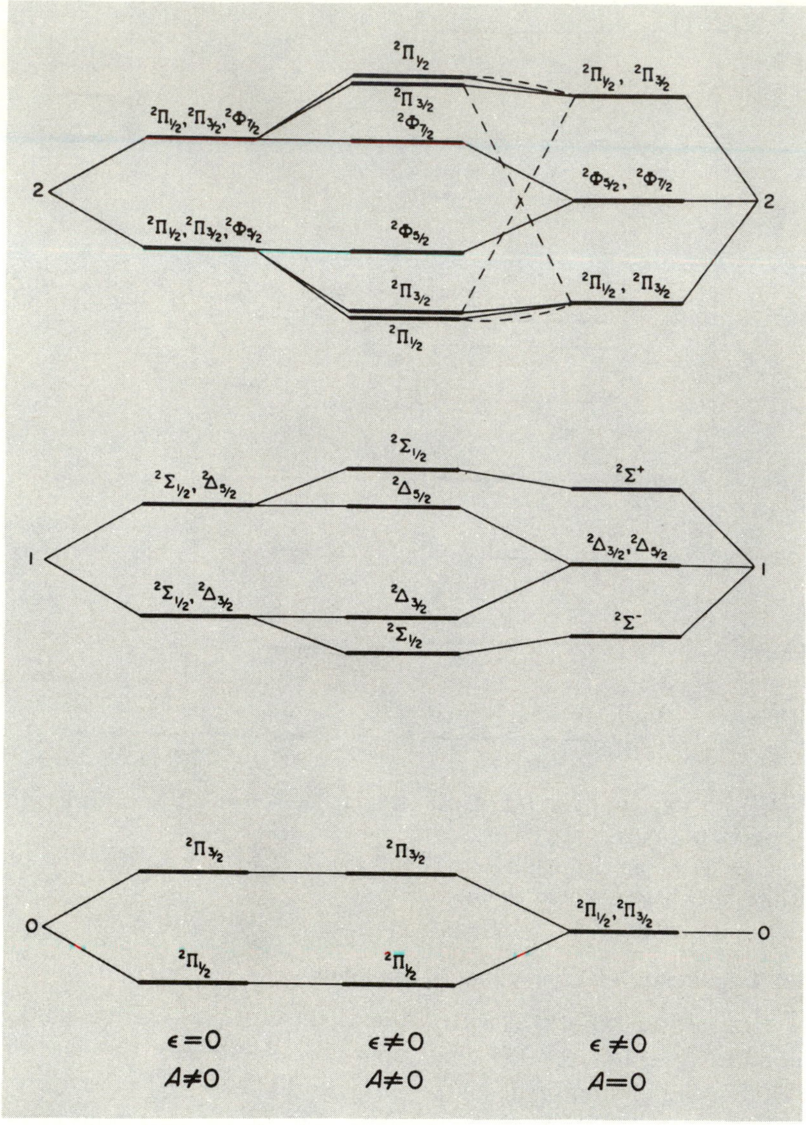

Fig. 60. Correlation of the vibronic levels in a $^2\Pi$ electronic state for zero vibronic (left) and zero spin-orbit (right) interaction with those for which both interactions are nonzero (center) (from *MM* III, p. 35).

Energy levels that coincide in a given approximation are shown as one level with the term symbols indicated.

action are comparable in magnitude—that is, when A and $\omega_2\epsilon$ are of the same order—the result is not simply a superposition of the two effects; there are considerable changes, as indicated in the center of Fig. 60. For example, some of the doublet splittings are increased compared to pure spin-orbit interaction. Detailed formulae developed by Pople (111) and Hougen (73) may be found in *MM* III, p. 36.

As the spin-orbit coupling increases from right to left in Fig. 60, the $^2\Sigma^+$ and $^2\Sigma^-$ vibronic levels show an increasing spin splitting and behave more like $\frac{1}{2}$ states [Hund's case (c), see *MM* I, p. 236]. They are designated $^2\Sigma_{1/2}$ in Fig. 60, since the symmetry property implied in the designation Σ^+ or Σ^- loses its meaning. At the same time there are changes of the rotational constants B in these levels. Detailed formulae may be found in *MM* III, p. 77f. The large spin splitting in $^2\Sigma$ vibronic states of $^2\Pi$ electronic states has been observed for NCO, BO_2, and similar free radicals, for which in fact Renner-Teller splittings were established long before this was done for singlet states.

Relations similar to $^2\Pi$ apply to $^3\Pi$ electronic states but will not be discussed here in detail (see *MM* III, pp. 37 and 82).

D. TRANSITIONS, EXAMPLES

(1) Rotation and rotation-vibration spectra

The rotation spectra of linear polyatomic radicals are entirely similar to those of diatomic molecules (see p. 53f.) and need no further discussion. These spectra occur in the microwave region, but up to now only one such spectrum has been observed for a free radical: NCO by Saito (121). Raman rotation spectra have not been observed for any radicals, but electron spin-resonance spectra have been obtained. No examples of spin-reorientation spectra of linear polyatomic molecules have been found.

Rotation-vibration spectra of linear polyatomic radicals are, of course, very similar to those of stable linear molecules (see *MM* II, chap. IV) as long as their ground states are Σ electronic states. In that case, for symmetrical molecules, Σ_u–Σ_g and Π_u–Σ_g vibrational transitions are expected in the infrared with rotational structures entirely similar to Σ–Σ and Π–Σ electronic bands of diatomic radicals. For symmetrical linear XY_2 molecules only ν_2 and ν_3 (see Fig. 53) are infrared-active. For unsymmetrical molecules the subscripts *g* and *u* must be dropped and all vibrations are infrared-active. No such spectra of radicals have as yet been observed in the gaseous phase, but in a solid matrix at very low temperature infrared fundamentals of a number of free radicals have been observed, especially by Milligan and Jacox. Naturally, under these conditions no rotational structure is resolved.

Linear polyatomic radicals that have a Π electronic ground state are expected to show a peculiarity in their infrared spectra not shown by stable molecules: since, because of vibronic interaction, the state in which a bending vibration is singly excited will be split into three sublevels Σ^+, Δ, Σ^- (see Fig. 55), the infrared bands corresponding to this bending vibration will be split into three subbands, since all three sublevels can combine with the Π ground state. This splitting depends on the value of the Renner parameter ϵ and may be quite large, of the order of 100 cm^{-1}. In the absence, so far, of infrared spectra of free radicals in the gaseous phase one is dependent on matrix spectra for a confirmation of this prediction. In the only case in which a detailed investigation has been made, that of NCO [Milligan and Jacox (95)], only one of the three subbands, $^2\Sigma^+$–$^2\Pi$, has been found. This is puzzling, since in a zero approximation the three subbands should have the same intensity.

(2) Electronic transitions: selection rules

The selection rules for electronic transitions of linear polyatomic molecules are the same as for diatomic molecules (see p. 51). But there are now two additional causes for the breakdown of these selection rules—that is, for the occurrence of forbidden transitions:

(a) If the molecule is nonlinear in the excited state, certain quantum numbers or symmetries are not defined in this state and some of the selection rules do not apply. This case will be discussed in Chapter IV.

(b) If strong vibronic interactions exist in one (or both) of the two states involved in an electric-dipole-forbidden electronic transition, this interaction may cause certain vibronic transitions (but not the 0–0 band) to appear as electric dipole radiation.

In order to understand the second cause we need only consider the previous treatment of transition moments in diatomic molecules, which can be applied to linear polyatomic molecules if we realize that ψ_v is now a function of all the $(3N - 5)$ normal coordinates. The transformation of the previous general expression (96) for the transition moment into (99) is valid only as long as ψ_{ev} can be resolved into the product (114). If vibronic interaction is not negligible, the product representation is not valid, and we must use the general formula (96) rather than (99) in order to obtain the transition moment. In that case, even if $R_{e'e''} = 0$—that is, even if we have a forbidden electronic transition—the vibronic transition moment $R_{e'v'e''v''}$ may be different from 0. This will happen when the species of ψ_{ev} is different from that of ψ_e. Thus for certain vibrational transitions a nonvanishing dipole transition moment will arise. For example a ${}^1\Sigma_g^+ - {}^1\Sigma_g^+$ electronic transition is electric-dipole forbidden ($R_{e'e''} = 0$), but if in the upper or lower state (not both) a σ_u^+ vibration (ν_3 in XY_2) is singly excited, the resulting *vibronic* transition would be of the type $\Sigma_u^+ - \Sigma_g^+$ and is electric-dipole allowed according to Eq. (96). No examples of forbidden transitions produced by vibronic interactions have as yet been observed for linear polyatomic radicals. For stable molecules several cases are known—for example, in dicyanoacetylene [Miller and Hannan (93)].

(3) Vibrational structure of electronic transitions

As Teller (69) first recognized, there are, unlike the situation for diatomic molecules (p. 52), specific selection rules for the vibrational transitions in the band systems of a polyatomic molecule, provided that the molecule has symmetry. Just as for diatomic molecules in an allowed electronic transition (that is, one with $R_{e'e''} \neq 0$) the vibrational transitions are determined by the value of the overlap integral

$$\int \psi_{v'}{}^{*}\psi_{v''}\, d\tau_{v}, \tag{121}$$

where $\psi_{v'}$ and $\psi_{v''}$ are the vibrational eigenfunctions of the upper and lower state, respectively. For the integral (121) to be different from 0—that is, for the particular vibrational transition to be allowed—the product

$$\psi_{v'}{}^{*}\psi_{v''} \textit{ must be totally symmetric.} \tag{122}$$

Totally symmetric means symmetric with respect to all symmetry operations permitted by the point group to which the molecule belongs. If $\psi_{v'}{}^{*}\psi_{v''}$ were antisymmetric with respect to any symmetry operation, then the value of the integral (121) would change from a positive to a negative value for a simple coordinate transformation. On the other hand, the value of a definite integral cannot depend on any coordinate transformation, and therefore the integral (121) in this case must be zero. Only if there are no symmetry operations for which $\psi_{v'}{}^{*}\psi_{v''}$ is antisymmetric —that is, only if $\psi_{v'}{}^{*}\psi_{v''}$ is totally symmetric can the value of the integral (121) be different from zero.

For degenerate vibrational levels, appropriate linear combinations of the mutually degenerate eigenfunctions must be used. In either case, in order to fulfill the condition (122) the two vibrational levels must have the same vibrational species. Since the lowest vibrational level of the ground state of the molecule is always totally symmetric, we see that on the basis of this rule, in cold absorption only totally symmetric vibrational levels of the excited state can be reached.

If a molecule has a single totally symmetric vibration, as does a linear symmetric triatomic molecule or radical such as C_3 or NCN or BO_2, we expect to find a single progression in this vibration which is similar in all respects to a vibrational progression of a diatomic molecule (see p. 67). However, if there are two or more totally symmetric vibrations, then we obtain not only two or more progressions in these vibrations, but also all combinations of them. This is illustrated for the case of two totally symmetric vibrations in Fig. 61 in an energy-level diagram. Only the transitions from the lowest vibrational level of

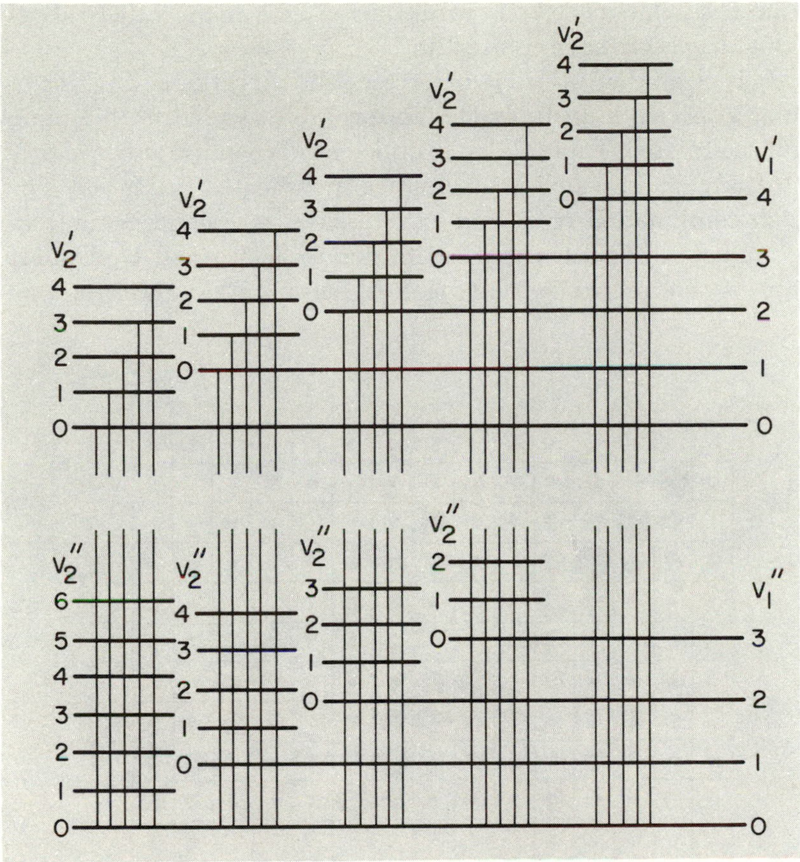

Fig. 61. Energy-level diagram showing the origin of progressions in cold absorption for a molecule with two totally symmetric vibrations ν_1 and ν_2 (from *MM* III, p. 144).

Only transitions from the lowest vibrational level of the ground state are shown. Totally symmetric levels corresponding to nontotally symmetric vibrations are not shown.

the ground state are shown (corresponding to cold absorption). It is seen that the vibrational structure of the spectrum in this simple case, of which a linear unsymmetric triatomic molecule XYZ would be an example, is much more complicated than in a diatomic molecule, for which only the first progression at the left would occur. For absorption at higher temperature, of course, similar transitions would arise from many other vibrational levels of the ground state.

It is clear that in order to present the vibrational analysis of such a spectrum a simple Deslandres table as used for diatomic molecules (see Table 7) is insufficient. We must use a *double Deslandres table*, which is schematically indicated in Fig. 62. For each combination of v_1' and v_1'' we have a Deslandres table in v_2' and v_2''. The intensity distribution in each small Deslandres table is determined by the *Franck-Condon principle* in the same way

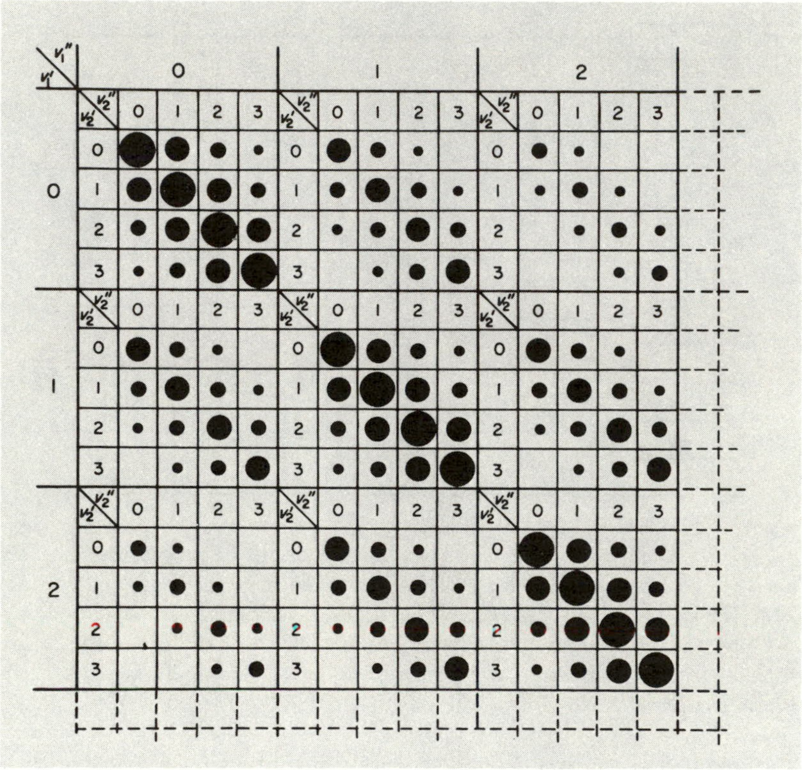

Fig. 62. "Double" Deslandres table of a band system of a polyatomic molecule involving two vibrations (from *MM* III, p. 145).

The size of the black dots is intended to indicate the intensity of the particular transition, assuming only a slight change of molecular dimensions in the transition and assuming nearly equal populations of the vibrational levels in the initial state.

as for diatomic molecules (see p. 69f.), and the same applies to the large Deslandres table.

If the two vibrational frequencies in the upper state are very similar in magnitude to those in the lower state, we find, as in diatomic molecules, the formation of *sequences*—that is, groups of bands that have the same value of Δv_1 and Δv_2. In such a case the strongest bands are those in the diagonals in each of the small Deslandres tables as well as in the diagonal of the large Deslandres table. The intensities are qualitatively indicated in Fig. 62 by

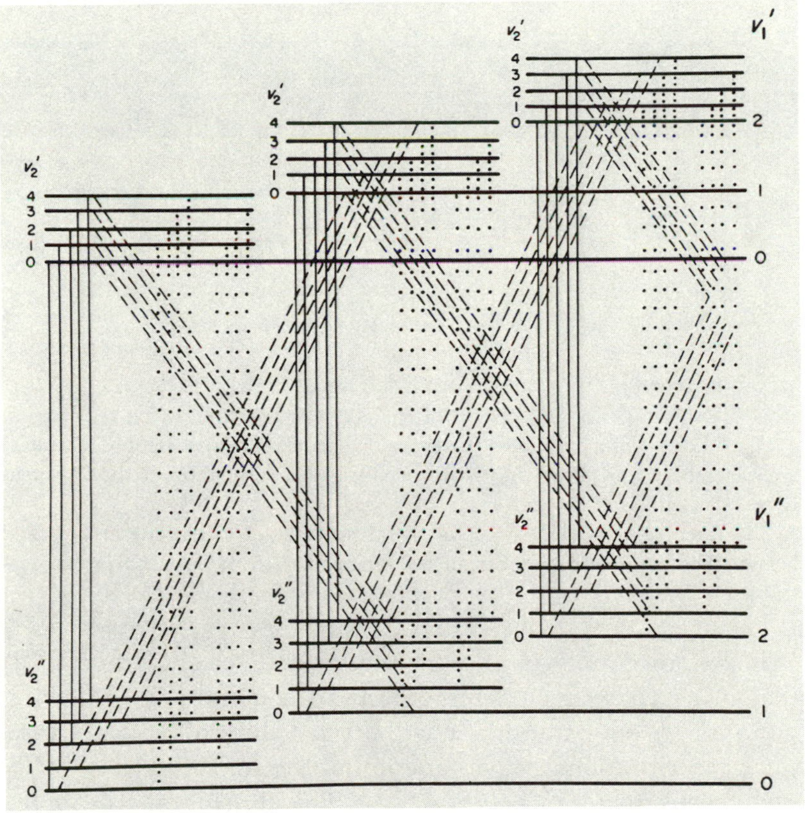

Fig. 63. Energy-level diagram showing the origin of sequences in a band system assuming that only two vibrations are active (from *MM* III, p. 147). The full-line transitions correspond to $\Delta v_1 = 0$, $\Delta v_2 = 0$, the broken-line transitions to $\Delta v_1 = \pm 1$, $\Delta v_2 = 0$, and the dotted transitions to $\Delta v_1 = 0$, $\Delta v_2 = \pm 1$.

the size of the black dots representing the individual transitions. Both in the small tables and in the large table the intensity decreases very rapidly as one goes away from the diagonal. The transitions of the sequences are indicated in an energy-level diagram in Fig. 63 for the case of two totally symmetric vibrations. A striking example of sequence structure is provided by the absorption and emission spectrum of the NCN free radical, which is shown in Fig. 64. Here only sequences with $\Delta v_i = 0$ are observed, and they occur in all three fundamentals.

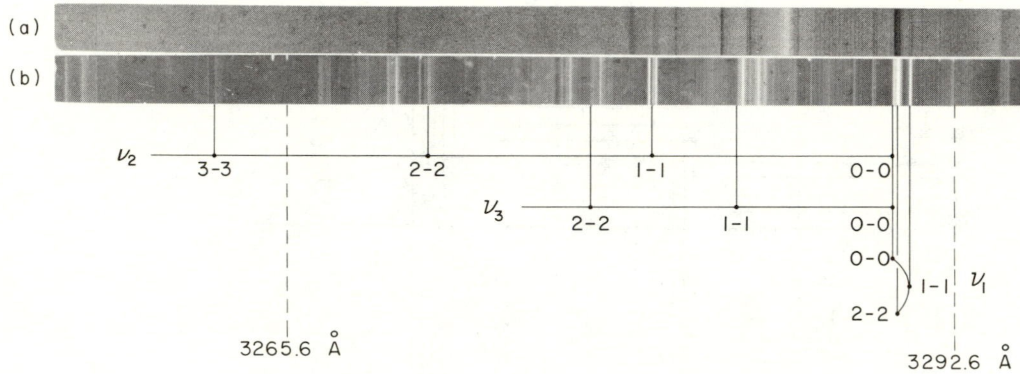

Fig. 64. (a) Absorption and (b) emission spectrum of the free NCN radical after Herzberg and Travis (70). [Reproduced by permission of the National Research Council of Canada from the *Canadian Journal of Physics.*]

The three sequences $\Delta v_i = 0$ in the three normal vibrations are marked below; they are more apparent in the emission than in the absorption spectrum.

It is immediately clear from the general selection rule (122) that in a progression corresponding to a nontotally symmetric vibration ν_k only transitions with even values of Δv_k can arise, since the vibrational wave function of the upper state is totally symmetric only for $v_k = 0, 2, 4, \ldots$. This conclusion also applies to the degenerate bending vibrations of linear molecules. As can be seen immediately from the diagram in Fig. 54, only for even v_k do Σ^+ (that is, totally symmetric) vibrational levels arise. The degenerate bending vibrations of linear molecules are characterized by the quantum number l_k, and for this quantum number according to the rule (122) we must have the selection rule

$$\Delta l_k = 0. \tag{123}$$

Figure 65 shows a double Deslandres table for a case in which one totally symmetric and one nontotally symmetric vibration are excited. On account of the restriction to even values of Δv_k, half the little squares are empty compared to Fig. 62.

There is another important difference of the intensity distribution in progressions in nontotally symmetric vibrations compared to progressions in totally symmetric vibrations. In the former, the intensity always decreases very rapidly from the first

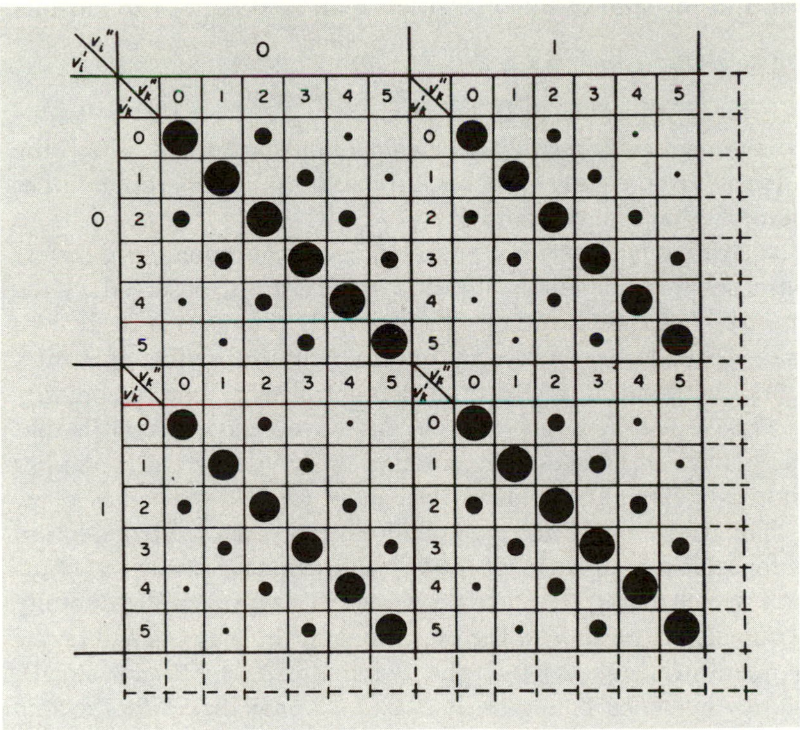

Fig. 65. Form of "double" Deslandres table for one symmetric (ν_i) and one antisymmetric (ν_k) vibration (from *MM* III, p. 152).

Compare Fig. 62. Note the absence of odd sequences in v_k.

band (with $\Delta v_k = 0$) on. This behavior is indicated by the size of the spots in Fig. 65, although in most practical cases the decrease is more rapid than shown in the diagram. The reason for the rapid decrease is the fact that the potential functions of nontotally symmetric vibrations are always even functions of the normal coordinate ξ_k and have their minima at the same value of this coordinate—namely, $\xi_k = 0$. In other words, excitation of a nontotally symmetric vibration corresponds to the diatomic case in which the potential functions of upper and lower state have their minima at the same place: therefore, only if the vibrational frequency of the upper state is markedly different from that of the lower state will transitions with $\Delta v_k \neq 0$ have non-negligible intensities. One finds for the intensity ratio of the 0–0 band to the sum of the intensities of all v_k–0 bands the formula

$$\frac{I_{0\text{-}0}}{\sum\limits_{v_k} I_{v_k\text{-}0}} = \frac{\sqrt{\omega_k'\omega_k''}}{\frac{1}{2}(\omega_k' + \omega_k'')}. \tag{124}$$

It is immediately seen from this formula that for $\omega_k' = \omega_k''$ the intensity ratio equals 1; that is, all bands of the progression other than 0–0 have zero intensity.

A striking example of a progression in a nontotally symmetric vibration (the bending vibration ν_2) has been observed in the spectrum of the C_3 radical, in which the bands 0–0, 0–2, 0–4, and 0–6 have been observed with rapidly decreasing intensity, while the bands 0–1, 0–3, and 0–5 are absent. The reason why in this case fairly large values of Δv_k have been observed is that the change of vibrational frequency is very large: in the upper state $\omega_2 = 308$, in the lower state $\omega_2 = 63$.

Unlike the rapid decrease of the intensity in a progression in a nontotally symmetric vibration, in a sequence of such a vibration with $\Delta v_k = 0$ the intensity varies in a way similar to totally symmetric vibrations: if the vibrational frequency is small or the temperature high (so that the Boltzmann factor is not small), higher sequence bands (1–1, 2–2, . . .) may have appreciable intensities; indeed, for degenerate vibrations the intensities of these "hot" bands are larger than for nondegenerate vibrations because of the greater statistical weight.

Further special considerations are necessary when one or both of the electronic states are degenerate and if vibronic interaction is not very small. In Fig. 66 the vibrational transitions connected with the bending vibration are shown for a $^1\Pi$–$^1\Sigma$ electronic transition of a linear triatomic molecule (a) without and (b) with vibronic interaction. The principal difference between the two cases is that the 1–1 and 2–2 transitions consist in the second case of three and five components, respectively, because of the

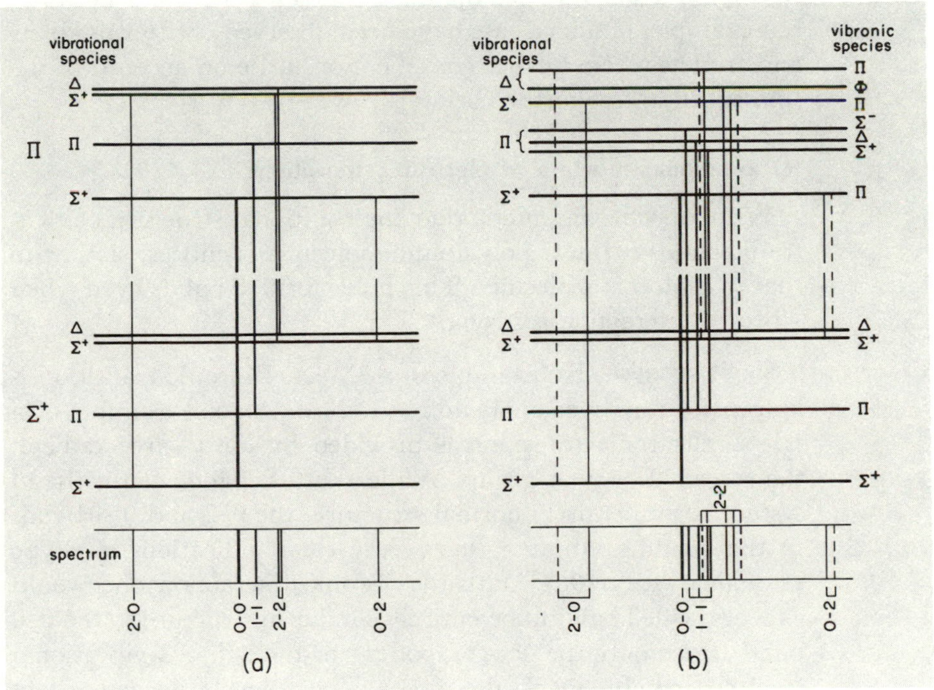

Fig. 66. Vibrational transitions in the bending vibration for a $^1\Pi$–$^1\Sigma$ electronic transition of a linear (triatomic) molecule (a) without and (b) with vibronic (Renner-Teller) splitting (from *MM* III, p. 159).

Schematic spectra are shown below the energy-level diagrams. Relative intensities are indicated by the weights of the lines representing the transitions. Transitions occurring only when vibronic interaction is introduced are represented by broken lines.

vibronic splittings in the $^1\Pi$ state. If the Renner parameter ϵ in the upper state is large, the separation of these three or five components will be large. Such splittings sometimes make difficult the vibrational analysis of $^1\Pi$–$^1\Sigma$ transitions in linear polyatomic radicals. An excellent example of these splittings is again provided by the C_3 free radical (see Fig. 59). From the observed splittings of the component bands, the quantity $\epsilon\omega_2$ is directly obtained.

Naturally for a $^1\Sigma$–$^1\Pi$ transition the situation is very similar, but no such case has yet been observed. For a $^1\Pi$–$^1\Pi$ transition we obtain again three component bands for the 1–1 transition, but examples of such a case have been observed only for doublets and triplets where further complications arise on account of the spin-splitting (see further below).

(4) Rotational structure of electronic transitions

Without vibronic interaction the rotational structure of electronic bands of linear polyatomic molecules is entirely similar to that of diatomic molecules. This holds for 0–0 bands even when vibronic interaction is strong.

Singlet bands. No example of a $^1\Sigma$–$^1\Sigma$ electronic transition is known for any linear polyatomic radical. A good example of a $^1\Pi$–$^1\Sigma$ electronic transition is provided by the C_3 free radical, the so-called 4050 Å group. While the 0–0 band[1] of this band system has an entirely normal structure, the 0–2 and 0–4 bands in the bending vibration show very clear indications of l-type resonance (see p. 96). Without vibronic interaction one would expect the 0–2 band to be entirely similar in structure to the 0–0 band (since only the Σ^+ component of the $v_2'' = 2$ vibrational level can combine with the upper $v_2' = 0$ level); in actual fact this band shows two component bands: one corresponding to the

[1] We are using here, in accordance with common practice, a short hand notation for what should be called 000–000 band. For molecules with more than three normal vibrations the full notation quickly becomes rather cumbersome. The 0–2 and 0–4 bands mentioned later in this paragraph have the full designation 000–020 and 000–040 respectively.

Π–Σ vibronic component, and the other corresponding to one half of the Π–Δ component, which is almost as strong as the former on account of l-type resonance.

In the same spectrum the combined effect of Λ-type doubling and l-type doubling is clearly shown by the large q values (see p. 96) in those 2–0 and 4–0 component bands that have a Π vibronic upper state. As shown by Johns (79), the q values are expected to be simply the sums of the q values for Λ-type and l-type doubling ($q_\Lambda + q_l$), and this is approximately found to be the case.

Another example of a singlet transition is the $^1\Pi$–$^1\Delta$ transition of the NCN free radical [Kroto (84)]. Here again the 000–000 band has a normal structure while the 010–010 band consists of four fairly widely separated component bands of vibronic types $^1\Sigma_g^+$–$^1\Pi_u$, $^1\Delta_g$–$^1\Phi_u$, $^1\Delta_g$–$^1\Pi_u$, $^1\Sigma_g^-$–$^1\Pi_u$, of which all but the $^1\Delta_g$–$^1\Pi_u$ component have been observed.

Doublet bands. For doublet transitions, as for singlet transitions, the 0–0 bands have the same structure as for diatomic molecules. A large number of examples of $^2\Sigma$–$^2\Pi$ electronic transitions have been observed—for example, for NCO, N_3, BO_2, CO_2^+, and CNC (see *MM* III, p. 499f and p. 592f). In each case the 1–1 bands have been found to consist of three component bands fairly widely spaced, corresponding to the three vibronic sublevels of the $v_2 = 1$ vibrational level of the ground state. In Fig. 67, as an example, part of the absorption spectrum of NCO is shown in which the three components of the 1–1 band can readily be distinguished. The structure of the $^2\Pi$–$^2\Delta$ vibronic component of the 1–1 band is similar to that of the 0–0 band except for the doubling of all branches on account of the l-type doubling in the upper and K-type doubling in the lower state. However, the structures of the $^2\Pi$–$^2\Sigma^\pm$ components of the 1–1 band are very different from that of the 0–0 band, since the doublet splitting in the $^2\Pi$ vibronic upper state is quite small (since the electronic state is $^2\Sigma$), while it is fairly large for the lower vibronic components $^2\Sigma^+$ and $^2\Sigma^-$ (since the electronic state is $^2\Pi$). The splitting in these vibronic $^2\Sigma$ states, as in $^2\Sigma$ electronic states, varies linearly with N [see Eq. (51a)], but the

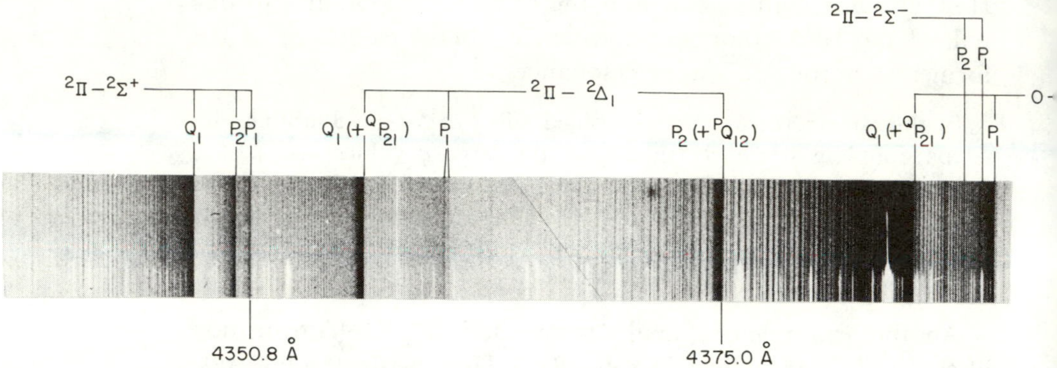

Fig. 67. Absorption spectrum of NCO showing the three components of the 010–010 bands of the $^2\Sigma$–$^2\Pi$ electronic transition after Dixon (31) (from *MM* III, p. 189).

The species symbols at the top designate the vibronic species (see Fig. 66). The $^2\Pi$–$^2\Sigma^-$ subband is very weak. The two strong heads at the right belong to the 0–0 band.

splitting constant γ is found to be quite large, of the same order as the rotational constant B (see p. 105).

$^2\Pi$–$^2\Pi$ electronic transitions have been observed in a number of triatomic free radicals such as NCO, BO_2, CCN, and CNC (see *MM* III, p. 498f and p. 591f). Here the 1–1 bands in the bending vibration have five subbands: $^2\Delta$–$^2\Delta$, $^2\Sigma^+$–$^2\Sigma^+$, $^2\Sigma^-$–$^2\Sigma^-$, $^2\Sigma^+$–$^2\Sigma^-$, and $^2\Sigma^-$–$^2\Sigma^+$, as shown in the energy-level diagram, Fig. 68. Only the first one of these vibronic subbands ($^2\Delta$–$^2\Delta$) has a normal (quasi-diatomic) structure with two fairly widely separated spin components, similar to the two components of the $^2\Pi$–$^2\Pi$ 0–0 band; the $^2\Sigma^+$–$^2\Sigma^+$ and $^2\Sigma^-$–$^2\Sigma^-$ subbands have spin-splittings that increase very rapidly with N, reflecting the large value of γ, quite unlike normal diatomic $^2\Sigma$–$^2\Sigma$ transitions; the $^2\Sigma^+$–$^2\Sigma^-$ and $^2\Sigma^-$–$^2\Sigma^+$ vibronic subbands which would be forbidden by the selection rule (60) do occur here because the doublet splitting is large and therefore the Σ^+, Σ^- property is no longer well defined (see *MM* I, p. 237). The structure of these subbands is unusual in that they contain only S-form, Q-form, and O-form branches ($\Delta N = +2, 0, -2$), somewhat similar to

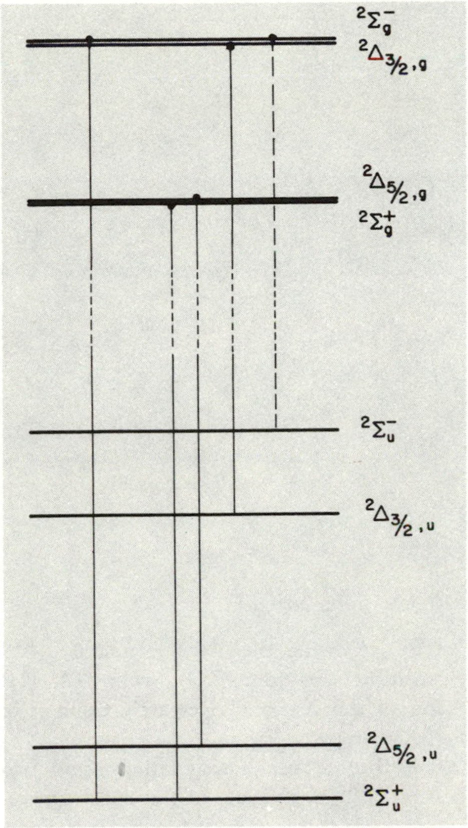

Fig. 68. Energy-level diagram for the subbands of the 010–010 band in a $^2\Pi_u$–$^2\Pi_g$ electronic transition of a linear XY_2 molecules.

The diagram is drawn approximately to scale for BO_2 after the data of Johns (78). Note that the $^2\Delta$–$^2\Delta$ vibronic subband consists of two fairly widely separated spin components, $^2\Delta_{3/2}$–$^2\Delta_{3/2}$ and $^2\Delta_{5/2}$–$^2\Delta_{5/2}$.

the $^1\Sigma^+$–$^3\Sigma^-$ bands of NF and SO discussed earlier. Figure 69 shows a spectrogram of the BO_2 $^2\Pi$–$^2\Pi$ system in which two of the components of the 1–1 transition are visible.

Triplet bands. $^3\Sigma$–$^3\Sigma$ electronic transitions in linear polyatomic radicals are entirely similar to those in diatomic molecules. Usually the triplet splitting in upper and lower state is small and, except at very high resolution, not resolved. An example is the CD_2 band at 1410 Å shown in Fig. 9. In this spectrum the odd lines are the strong ones, while in CH_2 the even lines are strong. This is the opposite of the situation in the $^1\Sigma_u^+$–$^1\Sigma_g^+$ absorption bands of H_2 and D_2 (the Lyman bands) and

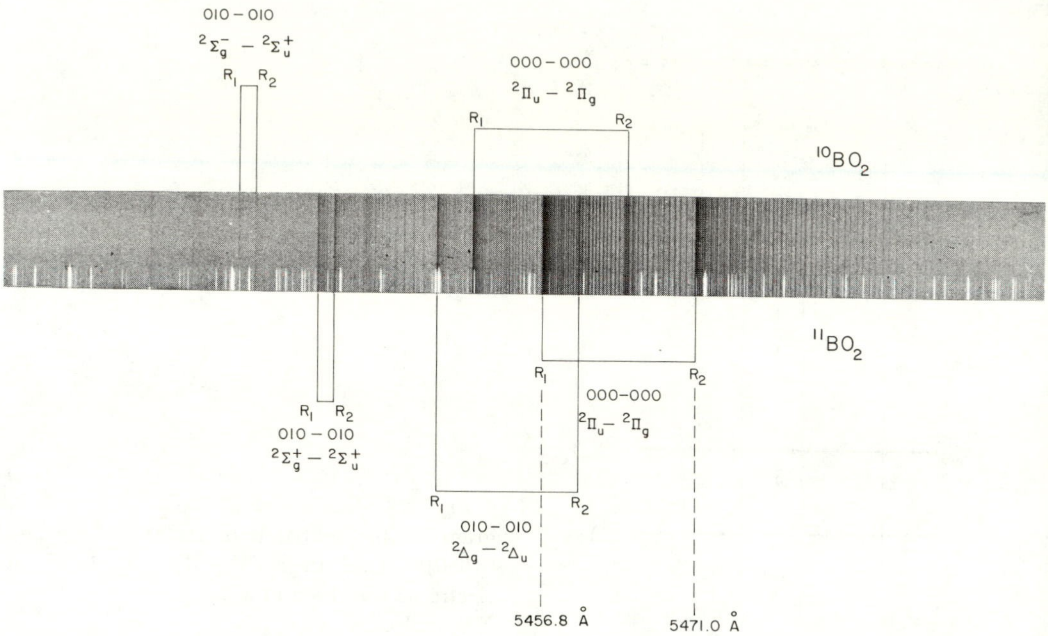

Fig. 69. Absorption spectrum of BO_2 showing the 000–000 and 010–010 bands of the $^2\Pi_u$–$^2\Pi_g$ electronic transition after Johns (78) (from *MM* III, p. 192). [Reproduced by permission of the National Research Council of Canada from the *Canadian Journal of Physics*.]

The subbands are identified by their vibronic species designations. Only two of the component bands of 010–010 of $^{11}BO_2$ are visible in this spectrogram.

shows that the lower state of the CD_2 band is not Σ_g^+, but must be either Σ_g^- or Σ_u^+. Since the predicted ground state of linear CH_2 (see below) is $^3\Sigma_g^-$, we conclude that the observed transition is $^3\Sigma_u^-$–$^3\Sigma_g^-$, even though the triplet splitting has not yet been resolved. The 1–1 bands in the bending vibration of $^3\Sigma$–$^3\Sigma$ electronic transitions would have the vibronic type $^3\Pi$–$^3\Pi$, but no examples have been observed.

The 0–0 band of a $^3\Pi$–$^3\Sigma$ electronic transition has three subbands, $^3\Pi_2$–$^3\Sigma$, $^3\Pi_1$–$^3\Sigma$, $^3\Pi_0$–$^3\Sigma$. If the $^3\Pi$ state belongs to Hund's case (a)—that is, if the spin splitting is large—each of the three subbands has nine branches, provided that the triplet splitting in the (lower) $^3\Sigma$ state is resolved. Most of the resulting 27 branches have been observed [Herzberg and Travis (70)] for the $^3\Pi_u$–$^3\Sigma_g^-$ transition of the NCN free radical near 3290 Å (see Fig. 64); the corresponding transition of the free CCO radical has recently been observed and analyzed by Devillers and Ramsay (28). The triplet nature of the ground state of these radicals is therefore directly and firmly established.

The 1–1 band in the bending vibration of a $^3\Pi$–$^3\Sigma$ electronic transition has three vibronic subbands: $^3\Delta$–$^3\Pi$, $^3\Sigma^+$–$^3\Pi$, and $^3\Sigma^-$–$^3\Pi$. Both for NCN and for CCO these subbands are observed to be widely separated, indicating strong vibronic interaction in the $^3\Pi$ electronic state. The first of the three subbands has a structure very similar to that of the 0–0 band except for the *l*-type doubling in the lower state, which causes all branches to be double compared to the 0–0 band. Only very few of the resulting 54 branches have been resolved in either NCN or CCO. The other two subbands show a triplet splitting that increases rapidly with N. This is not due to the splitting of the lower $^3\Pi$ vibronic state, since this state belongs to a $^3\Sigma$ electronic state in which the spin splitting is small; rather, the splitting is caused by vibronic interaction in the $^3\Pi$ upper electronic state. As we have seen, the spin-splitting constants λ and γ in such $^3\Sigma$ vibronic states are rather large, and as a result two sub-subbands arise for both $^3\Sigma^+$–$^3\Pi$ and $^3\Sigma^-$–$^3\Pi$, one corresponding to $^3\Sigma_1$, the other to $^3\Sigma_{0^-}$ and $^3\Sigma_{0^+}$, respectively.

No other type of triplet transition nor any quartet transitions have been observed.

E. ELECTRON CONFIGURATIONS

Dihydrides. Just as for diatomic hydrides, we obtain the electron configurations of triatomic dihydrides by comparison with the united atom. It is customary to use a simplified designation of the orbitals: $1\sigma_g$ means the lowest, $2\sigma_g$ the second lowest σ_g orbital; $1\sigma_u$ means the lowest σ_u orbital, and so on. On this basis Table 9 gives the electron configurations of the ground states and the first excited states of the dihydrides of the elements of the first period, assuming that they have a linear

Table 9. Ground States and First Excited States of Linear XH_2 Molecules as Derived from the Electron Configurations

Molecule	*Lowest Electron Configuration and Ground State*		*First Excited Electron Configuration and Resulting States*	
BeH_2	$(1\sigma_g)^2(2\sigma_g)^2(1\sigma_u)^2$	$^1\Sigma_g^+$	$\ldots(1\sigma_u)(1\pi_u)$	$^3\Pi_g$, $^1\Pi_g$
			$\ldots(1\sigma_u)(3\sigma_g)$	$^3\Sigma_u^+$, $^1\Sigma_u^+$
BH_2, CH_2^+	$(1\sigma_g)^2(2\sigma_g)^2(1\sigma_u)^2(1\pi_u)$	$^2\Pi_u$	$\ldots(1\sigma_u)^2(3\sigma_g)$	$^2\Sigma_g^+$
			$\ldots(1\sigma_u)(1\pi_u)^2$	$^4\Sigma_u^-$, $^2\Sigma_u^-$, $^2\Delta_u$, $^2\Sigma_u^+$
CH_2, NH_2^+	$(1\sigma_g)^2(2\sigma_g)^2(1\sigma_u)^2(1\pi_u)^2$	$^3\Sigma_g^-$, $^1\Delta_g$, $^1\Sigma_g^+$	$\ldots(1\sigma_u)^2(1\pi_u)(3\sigma_g)$	$^3\Pi_u$, $^1\Pi_u$
			$\ldots(1\sigma_u)(1\pi_u)^3$	$^3\Pi_g$, $^1\Pi_g$
NH_2, H_2O^+	$(1\sigma_g)^2(2\sigma_g)^2(1\sigma_u)^2(1\pi_u)^3$	$^2\Pi_u$	$\ldots(1\sigma_u)^2(1\pi_u)^2(3\sigma_g)$	$^4\Sigma_g^-$, $^2\Sigma_g^-$, $^2\Delta_g$, $^2\Sigma_g^+$
			$\ldots(1\sigma_u)(1\pi_u)^4$	$^2\Sigma_u^+$
H_2O	$(1\sigma_g)^2(2\sigma_g)^2(1\sigma_u)^2(1\pi_u)^4$	$^1\Sigma_g^+$	$\ldots(1\sigma_u)^2(1\pi_u)^3(3\sigma_g)$	$^3\Pi_u$, $^1\Pi_u$

structure; in fact, of those listed, only CH_2 is known to have a linear structure in its ground state; probably the BeH_2 radical also has a linear structure, but its spectrum has not yet been observed. The electron configurations of the other dihydrides, which are known to be nonlinear in their ground states, are given here under the assumption that they are linear for later comparison with the electron configurations of the nonlinear forms. In CH_2, since there are two π electrons, three low-lying states arise, ${}^3\Sigma_g^-$, ${}^1\Delta_g$, and ${}^1\Sigma_g^+$, of which the first is expected to be the lowest.

A Rydberg series of bands has been observed for CH_2 in the vacuum-ultraviolet region [Herzberg (56)]. The upper states probably correspond to the configurations.

$$\ldots (1\sigma_u)^2(1\pi_u)(nd\pi_g) \qquad {}^3\Sigma_u^+,\ {}^3\Sigma_u^-,\ {}^3\Delta_u,\quad {}^1\Sigma_u^+,\ {}^1\Sigma_u^-,\ {}^1\Delta_u. \tag{125}$$

There are, of course, many other Rydberg configurations, but they either do not combine, or combine only weakly, with the ground state and therefore have not been observed. The limit of the Rydberg series of CH_2 corresponds to an energy of 10.396 eV, which would be the ionization potential of CH_2 if in its ground state CH_2^+ were linear. However, the study of the spectrum of BH_2, which has the same number of electrons as CH_2^+, shows conclusively that BH_2 is nonlinear in its ground state, and this result strongly suggests that CH_2^+ in its ground state also is not linear, and therefore the value given is only an upper limit to the ionization potential of CH_2.[1]

Monohydrides. The electron configuration of triatomic monohydrides HXY are similar to those of the *united molecule* obtained when HX is replaced by its united atom. Table 10 lists the lowest electron configurations of a few monohydrides, assuming them to be linear. The ground state of HCC is expected to be ${}^2\Pi$, corresponding to the first excited state of the united molecule CN. The difference is due to a reversal in the order of the outermost σ and π orbitals (since the σ electrons, which are non-bonding in CN, are C—H bonding in HCC). The lowest electron configuration of HCF or HNO, just as that of the united

[1]See, however, the remark added on p. 16. It is now well established that both CH_2 and CH_2^+ in their ground states are nonlinear, with approximately the same angle (135°). Thus the ionization potential of CH_2 is 10.396 eV. [1988 note.]

Table 10. Lowest Electron Configurations of Linear HXY Molecules and Corresponding Electronic States[a]

Molecules	*Lowest Electron Configurations*	*Resulting States*
HCC	$\ldots\sigma^2\sigma^2\pi^3$	$^2\Pi$
HCN	$\ldots\sigma^2\sigma^2\pi^4$	$^1\Sigma^+$
HCO, HN_2	$\ldots\sigma^2\sigma^2\pi^4\pi$	$^2\Pi$
HCF, HNO	$\ldots\sigma^2\sigma^2\pi^4\pi^2$	$^3\Sigma^-$, $^1\Delta$, $^1\Sigma^+$
HNF	$\ldots\sigma^2\sigma^2\pi^4\pi^3$	$^2\Pi$
HOF	$\ldots\sigma^2\sigma^2\pi^4\pi^4$	$^1\Sigma^+$

[a] Of the molecules listed only HCN is known to be linear in its ground state.

molecules NF or O_2, has two equivalent π electrons, giving rise to the states $^3\Sigma^-$, $^1\Delta$, and $^1\Sigma^+$, the first of which is expected to be the lowest. Although spectra of HNO and HCF have been observed [Dalby (26), Bancroft, Hollas, and Ramsay (5), Merer and Travis (92)], the $^3\Sigma^-$ ground state of the linear conformation or its analogue in the bent conformation has not yet been found for either of them.

Nonhydrides. In order to derive the electron configurations of linear triatomic nonhydride molecules or radicals we must consider the correlation with the separated atoms in a way similar to the correlation diagrams for diatomic molecules, Figs. 16 and 17. Figure 70 gives a correlation diagram of the orbital energies for linear XY_2 molecules, showing the variation between large and small internuclear distances. The actual order of the orbital energies that we must use in order to determine the electron configurations of these molecules is somewhere near the center of this diagram. In Table 11 the lowest and the first excited electron configurations of a number of important linear triatomic free radicals are given, together with the resulting states. The observed states have been underlined. It is seen that many of the predicted states have been observed, and that the observed ground states are in accord with the predictions. Table 12 lists the rotational constants B_0 and the frequencies of the bending vibration in the ground states of the same radicals. Also, where available, the resulting internuclear distances are given. It is interesting to note that the frequency of the bending vibration rises from the extremely low value of 63 cm^{-1} for the ground state of C_3 to the value 667 cm^{-1} for that of CO_2. This increase appears to be connected with the filling of the $1\pi_g$ orbital.

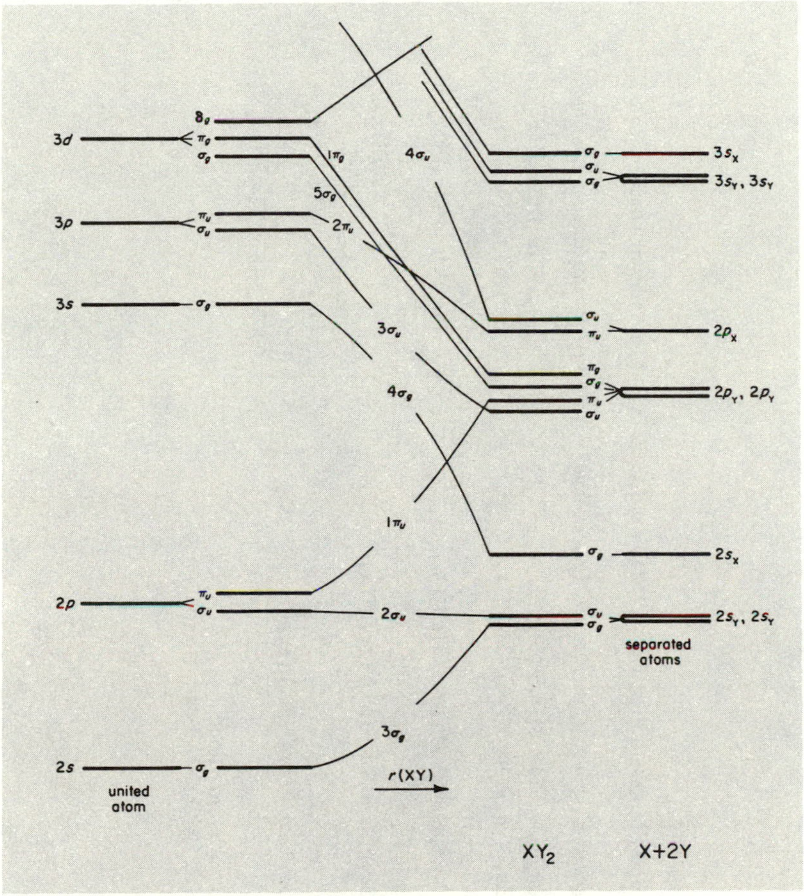

Fig. 70. Correlation of molecular orbitals between large and small internuclear distances in linear XY_2 molecules (from *MM* III, p. 314).

At the extreme left are the energies of the orbitals of the united atom, at the extreme right those of the separated atoms. It is assumed here that the energies of the $2p$ and $2s$ orbitals of X are higher than those of Y—that is, that the corresponding ionization potentials are smaller.

Table 11. Ground States and First Excited States of Linear XY_2 Molecules as Derived from the Electron Configurations[a]

Molecule	*Lowest Electron Configuration and Ground State*		*First Excited Electron Configuration and Resulting States*	
C_3	$(3\sigma_g)^2(2\sigma_u)^2(4\sigma_g)^2(1\pi_u)^4(3\sigma_u)^2$	$\underline{^1\Sigma_g^+}$	$\ldots(4\sigma_g)^2(1\pi_u)^4(3\sigma_u)(1\pi_g)$	$^3\Pi_u$, $\underline{^1\Pi_u}$
CNC, CCN[b]	$(3\sigma_g)^2(2\sigma_u)^2(4\sigma_g)^2(1\pi_u)^4(3\sigma_u)^2(1\pi_g)$	$\underline{^2\Pi_g}$	$\ldots(4\sigma_g)^2(1\pi_u)^4(3\sigma_u)(1\pi_g)^2$	$^4\Sigma_u^-$, $\underline{^2\Sigma_u^-}$, $\underline{^2\Sigma_u^+}$, $\underline{^2\Delta_u}$
NCN, NNC[b] CCO[b]	$(3\sigma_g)^2(2\sigma_u)^2(4\sigma_g)^2(1\pi_u)^4(3\sigma_u)^2(1\pi_g)^2$	$\underline{^3\Sigma_g^-}$, $\underline{^1\Delta_g}$, $^1\Sigma_g^+$	$\ldots(4\sigma_g)^2(1\pi_u)^4(3\sigma_u)(1\pi_g)^3$	$\underline{^3\Pi_u}$, $\underline{^1\Pi_u}$
			$\ldots(4\sigma_g)^2(1\pi_u)^3(3\sigma_u)^2(1\pi_g)^3$	$^3\Sigma_u^-$, $^3\Sigma_u^+$, $^3\Delta_u$, $^1\Sigma_u^-$, $^1\Sigma_u^+$, $^1\Delta_u$
BO_2, CO_2^+, N_3, NCO[b]	$(3\sigma_g)^2(2\sigma_u)^2(4\sigma_g)^2(3\sigma_u)^2(1\pi_u)^4(1\pi_g)^3$	$\underline{^2\Pi_g}$	$\ldots(4\sigma_g)^2(3\sigma_u)^2(1\pi_u)^3(1\pi_g)^4$	$\underline{^2\Pi_u}$
			$\ldots(4\sigma_g)^2(3\sigma_u)(1\pi_u)^4(1\pi_g)^4$	$\underline{^2\Sigma_u^+}$
CO_2	$(3\sigma_g)^2(2\sigma_u)^2(4\sigma_g)^2(3\sigma_u)^2(1\pi_u)^4(1\pi_g)^4$	$^1\Sigma_g^+$	$\ldots(4\sigma_g)^2(3\sigma_u)^2(1\pi_u)^4(1\pi_g)^3(2\pi_u)$	$^3\Sigma_u^-$, $^3\Sigma_u^+$, $^3\Delta_u$, $^1\Sigma_u^-$, $^1\Sigma_u^+$, $\underline{^1\Delta_u}$
			$\ldots(4\sigma_g)^2(3\sigma_u)^2(1\pi_u)^4(1\pi_g)^3(5\sigma_g)$	$^3\Pi_g$, $^1\Pi_g$

[a] Observed states are underlined.

[b] For CCN, NNC, CCO, NCO omit *g* and *u*.

Table 12. Rotational and Vibrational Constants in the Ground States of Linear Triatomic Nonhydrides

Molecule	*State*	ΔG (cm^{-1})	B_0 (cm^{-1})	r_0 (Å)	*N*
C_3	$^1\Sigma_g^+$	63.1	0.4305	1.277	12
CCN	$^2\Pi$	(325)	0.3981		13
CNC	$^2\Pi_g$	321	0.4535	1.245	
NCN	$^3\Sigma_g^-$	(423)	0.3968	1.232	14
NCO	$^2\Pi$	(539)	0.3894		
N_2O^+	$^2\Pi$	461.2	0.4116	1.155 1.185	
N_3	$^2\Pi_g$		0.4312	1.182	15
BO_2	$^2\Pi_g$	464	0.3292	1.265	
CO_2^+	$^2\Pi_g$		0.3804	1.177	
CO_2	$^1\Sigma_g^+$	667.4	0.3902	1.162	16

N = number of valence electrons.

IV. NONLINEAR POLYATOMIC RADICALS AND IONS

A. ELECTRONIC STATES

(1) Classification

A molecule, like any system of mass points, may have one or more *elements of symmetry:* a plane of symmetry, a center of symmetry, a p-fold axis of symmetry. To each symmetry element corresponds a *symmetry operation:* reflection at a plane of symmetry or at a center of symmetry or a rotation by $360°/p$ about an axis of symmetry. A linear molecule has an infinite number of elements of symmetry (every plane through the internuclear axis is a plane of symmetry) and therefore an infinite number of symmetry types (species) of molecular wave functions (Σ^+, Σ^-, Π, Δ, . . .), but a nonlinear molecule has a finite, usually small, number of elements of symmetry and as a consequence a finite number of symmetry types.

Only certain combinations of symmetry elements in any rigid system of points are possible. They are called *point groups.* A system of points with only one plane of symmetry (and no other elements of symmetry) belongs to the point group usually called $\boldsymbol{C_s}$. An unsymmetric nonlinear triatomic molecule XYZ has this symmetry. Since the potential energy for the electrons as well as the nuclei of such a system remains unchanged upon reflection at the plane of symmetry (usually called σ), all molecular wave functions must be either symmetric or antisymmetric with respect to this plane. Thus, we have only two species—that is, two kinds of electronic states—which are called A' and A''.

A symmetric nonlinear XY_2 molecule has two planes[1] of symmetry σ_v, and as a consequence also a twofold axis of symmetry C_2. It is said to belong to the point group $\boldsymbol{C_{2v}}$. This point group has four different species corresponding to the four ways of

[1] It is assumed by convention that the axis of symmetry is set up vertically; therefore the planes of symmetry are called *vertical planes*, σ_v.

assigning plus and minus signs to the two reflections at the two planes of symmetry. These species, called A_1, A_2, B_1, B_2, are defined by the "*characters*" given in Table 13(a).

Table 13. Symmetry Types (Species) and Characters for the Point Groups $\boldsymbol{C_{2v}}$ and $\boldsymbol{C_{2h}}$

(a)

$\boldsymbol{C_{2v}}$	I	$C_2(z)$	$\sigma_v(xz)$	$\sigma_v(yz)$
A_1	+1	+1	+1	+1
A_2	+1	+1	−1	−1
B_1	+1	−1	+1	−1
B_2	+1	−1	−1	+1

(b)

$\boldsymbol{C_{2h}}$	I	$C_2(z)$	$\sigma_h(xy)$	i
A_g	+1	+1	+1	+1
A_u	+1	+1	−1	−1
B_g	+1	−1	−1	+1
B_u	+1	−1	+1	−1

The four columns show in each case the behaviour for the four symmetry operations; I is the identity operation; $C_2(z)$ is a rotation by 180° about the twofold axis assumed to be the z axis; $\sigma_v(xz)$, $\sigma_v(yz)$ are reflections at the "vertical" planes xz and yz, $\sigma_h(xy)$ at the "horizontal" plane xy; i is the inversion (reflection at the center of symmetry). The behaviour with respect to $C_2(z)$ for $\boldsymbol{C_{2v}}$ follows from that for $\sigma_v(xz)$ and $\sigma_v(yz)$ simply by multiplication of the corresponding characters; similarly for $\boldsymbol{C_{2h}}$ the characters for $C_2(z)$ follow from those for $\sigma_h(xy)$ and i.

A nonlinear four-atomic molecule X_2Y_2 may have a planar *cis* structure:

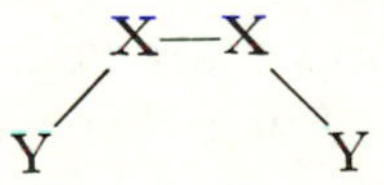

or a planar *trans* structure:

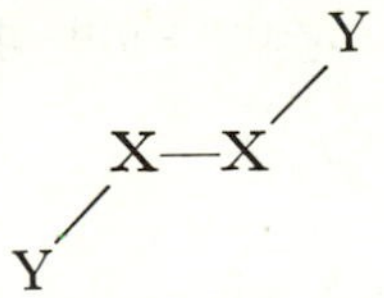

In the first case it belongs to the point group $\boldsymbol{C_{2v}}$ just discussed, with electronic states of species A_1, A_2, B_1, B_2. In the second case the twofold axis is perpendicular to the plane of the molecule, which is now the only plane of symmetry, called σ_h ("horizontal" plane). The presence of C_2 and σ_h implies that there is also a center of symmetry i. This combination of symmetry elements is called $\boldsymbol{C}_{2h}$. The point group $\boldsymbol{C}_{2h}$, like $\boldsymbol{C}_{2v}$, has four species, designated A_g, A_u, B_g, B_u, where the subscripts g and u refer to symmetry and antisymmetry, respectively, with regard to the center of symmetry, while A and B as before denote symmetry or antisymmetry with respect to the operation C_2. The characters of the four species with respect to all symmetry elements are given in Table 13(b).

If the X_2Y_2 molecule is nonplanar, the point group is $\boldsymbol{C}_2$, which has only two species, A and B, corresponding to symmetry and antisymmetry with respect to rotation by 180° about the axis of symmetry C_2.

For a nonplanar symmetrical XY_3 molecule the point group is $\boldsymbol{C}_{3v}$. In this case we have a threefold axis of symmetry C_3 and three ("vertical") planes of symmetry σ_v through this axis. Because of the presence of the threefold axis we have one doubly degenerate species, E, which in some respects is similar to the species Π of linear molecules. When the symmetry operation C_3 is carried out, the wave function ψ does not simply remain unchanged or change sign, but goes over into a different function; however, all the functions obtained by the various symmetry operations can be represented as linear combinations of two functions; that is, we have a double degeneracy. The other two species of point group $\boldsymbol{C}_{3v}$ are nondegenerate; their symmetry properties (characters), together with those of E, are given in Table 14. For the degenerate species the characters are the sums of the diagonal terms in the matrix corresponding to the transformation representing the symmetry operation considered.

Table 14. Symmetry Types (Species) and Characters for the Point Group $\boldsymbol{C}_{3v}$

	I	$2C_3(z)$	$3\sigma_v$
A_1	+1	+1	+1
A_2	+1	+1	−1
E	+2	−1	0

For planar XY_3 molecules we have an additional plane of symmetry σ_h perpendicular to the symmetry axis, which entails several other elements of symmetry. The resulting eigenfunctions can be either symmetric or antisymmetric with respect to the plane σ_h, and this behaviour is indicated by a prime or double prime added to the species of $\boldsymbol{C}_{3v}$ (Table 14). Thus, we have for this point group, which is called $\boldsymbol{D}_{3h}$, the species A_1', A_1'', A_2', A_2'', E', E''.

For higherfold axes more and more degenerate species arise, called E_1, E_2, E_3, . . . , which are in many respects similar to Π, Δ, Φ, . . . of linear molecules except that their number is finite (only E_1 and E_2 for five- and sixfold axes, E_1, E_2, E_3 for seven- and eightfold axes, and so on).

The electron spin causes a *multiplicity* $2S + 1$ of the electronic states. If spin-orbit coupling is small, the effect of the spin can be taken into account in much the same way as in Hund's case (b) in diatomic and linear polyatomic molecules (see p. 44f.). This is the usual situation for nonlinear molecules containing only light atoms. For large spin-orbit coupling rather more complicated considerations are required, which in the case of even multiplicity require the use of the so-called double groups. A detailed discussion of such cases may be found in *MM* III.

(2) Electron Configurations

For each electron, just as for diatomic and linear polyatomic molecules, there are orbital eigenfunctions (or, for short, orbitals) corresponding to the motion of the electron considered in the field of the fixed nuclei and the average field of the other electrons. As for linear molecules, to designate the orbitals we use small letters corresponding to the capital letters designating the species. In other words, we distinguish a' from a'' orbitals for molecules belonging to point group $\boldsymbol{C}_s$; a_1, a_2, b_1, b_2 orbitals for $\boldsymbol{C}_{2v}$ molecules; a_g, a_u, b_g, b_u orbitals for $\boldsymbol{C}_{2h}$ molecules, and so on.

Approximate order of orbital energies. In a rough approximation one can obtain the approximate order of the energies of the various orbitals from the correlation between the united atom or molecule and the separated atoms, in much the same way as was described earlier for diatomic and linear polyatomic molecules. We must, however, realize that for each type of molecule there is a different correlation. Let us consider as an example the case of nonlinear XH_2. In Fig. 71 at the extreme left the orbital energies of the united atom are plotted, and at the extreme right

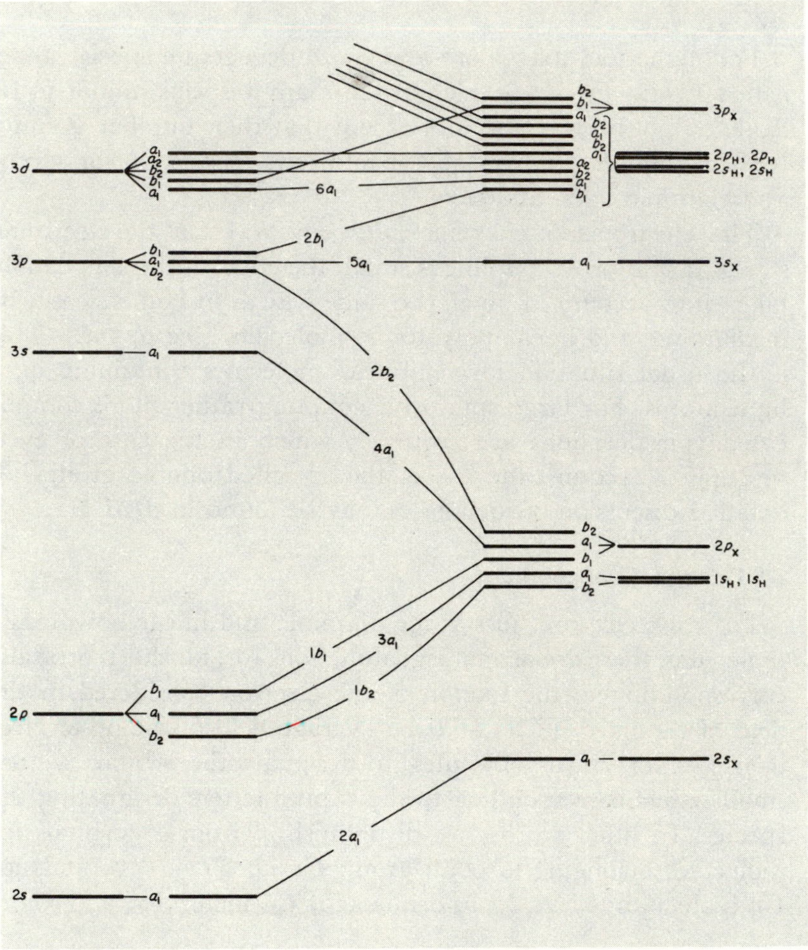

Fig. 71. Correlation of molecular orbitals between large and small internuclear distances in bent XH_2 molecules (from *MM* III, p. 317). See caption of Fig. 70.

those of the separated atoms. In the molecule, *s* orbitals of the united atom or of the separated atoms become a_1 orbitals; *p* orbitals split into three orbitals, a_1, b_1, and b_2; and *d* orbitals split into five orbitals, a_1, a_1, a_2, b_1, b_2. Since there are two hydrogen atoms, there are two orbitals of equal energy marked $1s_H$ at the extreme right in Fig. 71. These two orbitals, when the internuclear distance is reduced, yield two molecular orbitals of different energy, one symmetric and the other antisymmetric with respect to the plane of symmetry perpendicular to the molecular plane—that is, an a_1 and a b_2 orbital. In correlating between left and right in Fig. 71 we must connect the lowest a_1 on the right with the lowest a_1 on the left, the lowest b_1 on the right with the lowest b_1 on the left, and so on. Thus we obtain in the central part of Fig. 71 very roughly the order of the molecular orbitals of nonlinear XH_2.

If we compare the order of orbitals in Fig. 71 with the order or orbitals of linear XH_2 given earlier, we can plot a diagram

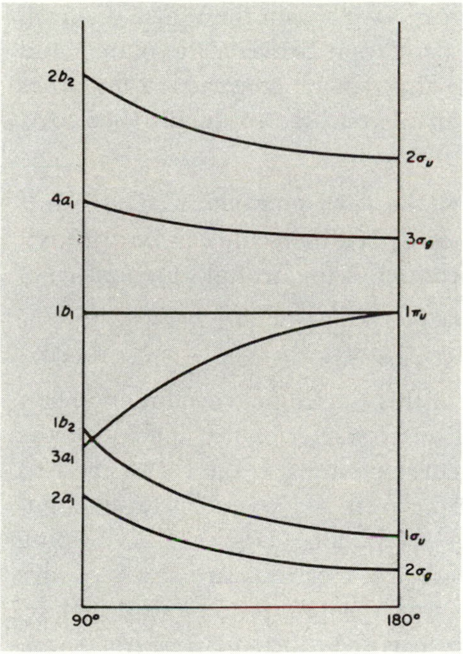

Fig. 72. Walsh diagram for XH_2 molecules (from *MM* III, p. 319).

The variation of orbital energies in going from a bent (90°) to a linear conformation is shown. The 1*s* orbital of X is not included.

that shows the correlation between linear and nonlinear XH_2 molecules. Such a diagram, first used by Walsh (135) and therefore often called a *Walsh diagram,* is shown in Fig. 72. In this diagram σ_g of linear XH_2 is correlated with a_1 of nonlinear XH_2 and similarly σ_u with b_2; the π orbitals split into two orbitals, one symmetric and the other antisymmetric with respect to the plane of the molecule. For π_u this gives a correlation with $a_1 + b_1$, and for π_g with $a_2 + b_2$.

As a second example, let us consider the correlation between planar and nonplanar XH_3. In Fig. 73 the correlation of the orbitals between large and small internuclear distance is shown for each of the two cases (similar to Fig. 71). Here it may be noted that a p orbital of the united atom or of the separated atoms splits into an a_1 and an e orbital in a $\boldsymbol{C}_{3v}$ molecule, and into an a_2'' and an e' orbital in a $\boldsymbol{D}_{3h}$ molecule. The three equivalent $1s_\mathrm{H}$ orbitals at large separation form a_1 and e (or a_1' and e') orbitals. The centers of Fig. 73(a) and (b) show roughly the order of orbitals in planar and nonplanar XH_3 molecules. By combination of these two diagrams we obtain the Walsh diagram of Fig. 74, which shows the correlation between the planar and nonplanar case. Similar diagrams can be constructed for other cases, but they will not be considered here in detail. (See *MM* III, p. 319.)

Molecular wave functions and the Pauli principle. The electronic molecular wave functions ψ_e of a molecule can be approximated very roughly by the product of the orbital functions ϕ_i of the individual electrons:

$$\psi_e = \phi_1\phi_2\phi_3 \ldots \tag{126}$$

The species (symmetry type) of this electronic wave function can be obtained as the so-called *direct product* of the species of the individual orbital functions (corresponding to the vector method for obtaining the states resulting from a given electron configuration of diatomic molecules—see p. 30f.). However, in forming this direct product, when *equivalent* electrons are involved, we must consider the restrictions introduced by the Pauli principle. For the resulting singlet states it is then the so-called *symmetrical*

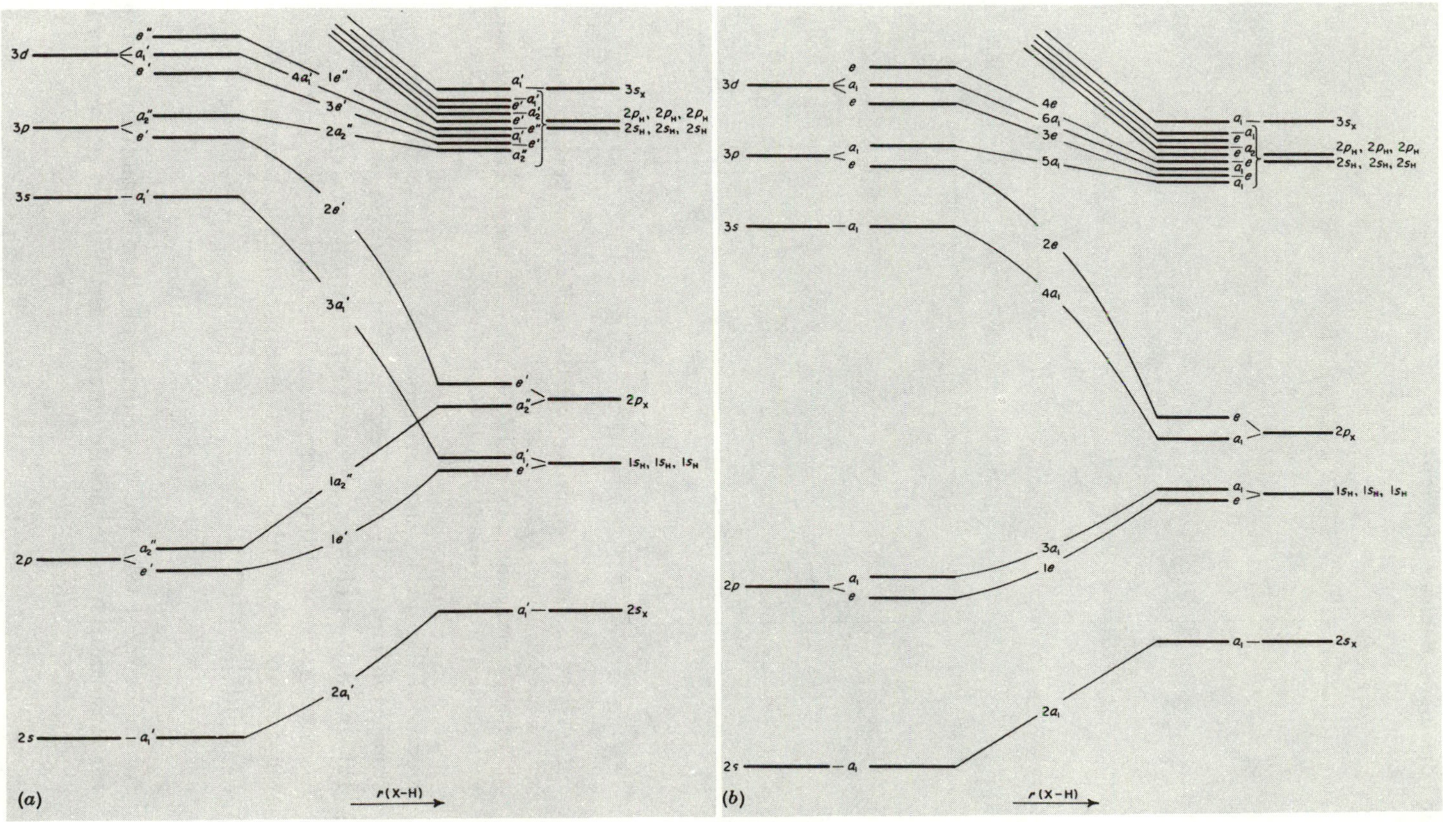

Fig. 73. Correlation of molecular orbitals between large and small internuclear distances (a) in planar XH_3 (point group $\boldsymbol{D}_{3h}$), (b) in non-planar XH_3 (point group $\boldsymbol{C}_{3v}$) (from *MM* III, p. 320).

See caption of Fig. 70. It is assumed that the ionization potential of X from the $2p$ orbital is somewhat less that of H from the $1s$ orbital.

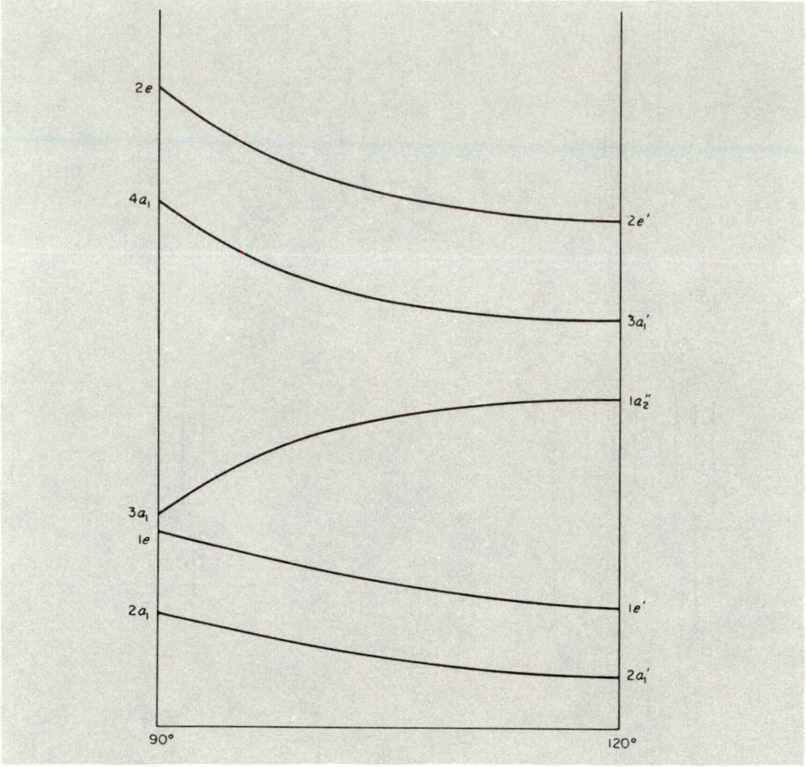

Fig. 74. Walsh diagram for the correlation of orbitals between nonplanar and planar XH_3 (from *MM* III, p. 321).

The angle on the abscissa is the HXH angle; 120° corresponds to the planar conformation. The 1*s* orbital of X is not included.

product of the species that matters, and for triplet states the *antisymmetrical product*. (For an explanation of these terms see *MM* III, p. 18, and elementary texts on group theory.)

The total spin of the molecule is obtained in the same way as for diatomic molecules—it is the sum of the spin vectors of the individual electrons,

$$\mathbf{S} = \sum \boldsymbol{s}_i. \tag{127}$$

Again we must take the Pauli principle into account in forming this resultant spin. A nondegenerate orbital can at most hold two electrons, and if it does, the two electrons must have antiparallel spins—that is, the total spin is zero; a doubly degenerate orbital cannot hold more than four electrons, and if it does, again the total spin is zero. This is true quite generally for *closed shells*—that is, whenever an orbital has the maximum number of electrons permitted by the Pauli principle.

Examples of the states resulting from various electron configurations with nonequivalent electrons are given for several point groups in Table 15, and for a number of configurations of

Table 15. Electronic States Resulting from Nonequivalent Electrons

Point Group	*Electron Configuration*	*Resulting States*
$\mathbf{C}_{2v}$	a_1	$^2A_1$
	a_1a_1	$^1A_1$, $^3A_1$
	a_1a_2	$^1A_2$, $^3A_2$
	b_1b_2	$^1A_2$, $^3A_2$
$\mathbf{C}_{3v}$	e	^2E
	a_1e	^1E, ^3E
	ee	$^1A_1$, $^1A_2$, ^1E, $^3A_1$, $^3A_2$, ^3E
$\mathbf{D}_{3h}$	$a_2''e'$	$^1E''$, $^3E''$
	$e'e''$	$^1A_1''$, $^1A_2''$, $^1E''$, $^3A_1''$, $^3A_2''$, $^3E''$
	$a_1''e'e''$	$^2A_1'(2)$, $^2A_2'(2)$, $^2E'(2)$, $^4A_1'$, $^4A_2'$, $^4E'$

equivalent electrons in Table 16. As for diatomic and linear polyatomic molecules, closed shells always give a totally symmetric species and $S = 0$—that is, a singlet state. A single electron in an orbital outside closed shells gives a state with a species equal to that of the orbital of the single electron, and this state will be a doublet state. For doubly degenerate orbitals, which can hold up to four electrons, a configuration of three electrons will give a state of the same species as a configuration of only one electron in that orbital. As indicated in Table 16, for two electrons in a doubly degenerate orbital three states result, which are analogous to the three states of the configuration π^2 of linear molecules.

Table 16. States Resulting from Equivalent Electrons

Point Group	*Electron Configuration*	*Resulting States*
$\mathbf{C}_{3v}$	a_2	$^2A_2$
	a_2^2	$^1A_1$
	e	^2E
	e^2	$^1A_1$, ^1E, $^3A_2$
	e^3	^2E
	e^4	$^1A_1$
$\mathbf{D}_{3h}$	e'	$^2E'$
	e'^2	$^1A_1'$, $^1E'$, $^3A_2'$
	e'^3	$^2E'$
	e'^4	$^1A_1'$
	e''	$^2E''$
	e''^2	$^1A_1'$, $^1E'$, $^3A_2'$
	e''^3	$^2E''$
	e''^4	$^1A_1'$
$\mathbf{C}_{4v}$	e	^2E
	e^2	$^1A_1$, $^1B_1$, $^1B_2$, $^3A_2$
	e^3	^2E
	e^4	$^1A_1$
$\mathbf{C}_{6v}$	e_1	$^2E_1$
	e_1^2	$^1A_1$, $^1E_2$, $^3A_2$
	e_1^3	$^2E_1$
	e_1^4	$^1A_1$
	e_2	$^2E_2$
	e_2^2	$^1A_1$, $^1E_2$, $^3A_2$
	e_2^3	$^2E_2$
	e_2^4	$^1A_1$
T_d	f_1	$^2F_1$
	f_1^2	$^1A_1$, ^1E, $^1F_2$, $^3F_1$
	f_1^3	^2E, $^2F_1$, $^2F_2$, $^4A_1$
	f_1^4	$^1A_1$, ^1E, $^1F_2$, $^3F_1$
	f_1^5	$^2F_1$
	f_1^6	$^1A_1$
	f_2	$^2F_2$
	f_2^2	$^1A_1$, ^1E, $^1F_2$, $^3F_1$
	f_2^3	^2E, $^2F_1$, $^2F_2$, $^4A_2$
	f_2^4	$^1A_1$, ^1E, $^1F_2$, $^3F_1$
	f_2^5	$^2F_2$
	f_2^6	$^1A_1$

Term manifold, stability, and geometrical structure. The term manifold of nonlinear molecules is obtained in the same way as previously described for diatomic molecules. First the electrons are assigned to their lowest possible orbitals as far as compatible with the Pauli principle. This yields the ground state of the molecule. Excited states are obtained by taking the outermost electron to the next highest orbital and then the next, and so on, or by taking an electron from one of the inner orbitals to the various unoccupied orbitals. As an example, in Table 17, the

Table 17. Ground States and First Excited States of Several Nonlinear XH_2 Molecules as Derived from the Electron Configurations

Molecule	*Lowest Electron Configuration and Ground State*		*First Excited Electron Configuration and Resulting States*	
BeH_2	$(1a_1)^2(2a_1)^2(1b_2)^2$	$^1A_1$	$\ldots(1b_2)(3a_1)$	$^3B_2$, $^1B_2$
BH_2	$(1a_1)^2(2a_1)^2(1b_2)^2(3a_1)$	$^2A_1$	$\ldots(1b_2)^2(1b_1)$	$^2B_1$
			$\ldots(1b_2)(3a_1)^2$	$^2B_2$
CH_2	$(1a_1)^2(2a_1)^2(1b_2)^2(3a_1)^2$	$^1A_1$	$\ldots(1b_2)^2(3a_1)(1b_1)$	$^3B_1$, $^1B_1$
NH_2	$(1a_1)^2(2a_1)^2(1b_2)^2(3a_1)^2(1b_1)$	$^2B_1$	$\ldots(1b_2)^2(3a_1)(1b_1)^2$	$^2A_1$
H_2O	$(1a_1)^2(2a_1)^2(1b_2)^2(3a_1)^2(1b_1)^2$	$^1A_1$	$\ldots(1b_2)^2(3a_1)^2(1b_1)(4a_1)$	$^3B_1$, $^1B_1$

lowest and the next to the lowest electron configurations and the corresponding resulting electronic states are given for the same series of XH_2 molecules for which in Table 9 the configurations were given assuming these molecules to be linear, but now assuming them to be nonlinear. Actually all the molecules listed other than BeH_2 (which has not been observed) are nonlinear in their ground states, but for BH_2 and CH_2 the bond angle is large (131° and 136° respectively). Note that $^3B_1$ forms the ground state of CH_2 (see p. 16). The states $^1A_1$ and $^1B_1$ of CH_2 are derived from the $^1\Delta$ state predicted for linear CH_2 (Table 9).

The stability of the various electronic states of the molecule is again determined, as in the case of diatomic molecules [see p. 39], by the number of bonding electrons—that is, electrons in orbitals that in diagrams such as Fig. 71 and Fig. 73 move downward on going from large to small internuclear distances. [For more details about this question see *MM* III, chap. III, sec. 3(b).]

We can obtain a rough prediction of the geometrical structure of the molecules by using Walsh diagrams. For example, from Fig. 72 we can readily see that for BeH_2 with four electrons outside the K shell, the ground state will be linear, because the energy curves of the two orbitals into which these four electrons will go have a minimum at an angle of 180°. The additional electron in BH_2 must go into the orbital $3a_1$, which according to Fig. 72 strongly favors a bond angle of 90° and is likely to overcompensate the action of the two other orbitals, which favor 180°. In agreement with this prediction the ground state of BH_2 has been found to have a bond angle of 131° [Herzberg and Johns (64)].

On the same basis, CH_2, with two electrons more than BeH_2, might be expected to have a bond angle even smaller than that of BH_2. Indeed, in the lowest singlet state (see Table 17) the bond angle is observed to be 102.4° [Herzberg and Johns (63)]. However, one of the two electrons can also be put into the other orbital ($1b_1$) arising from $1\pi_u$ of linear XH_2 (see Fig. 72). The $1b_1$ orbital favors neither the linear nor the bent conformation. Two states result from the configuration $3a_1 1b_1$, namely $^1B_1$ and $^3B_1$ (see Table 17), of which the latter is expected to be the lower according to Hund's rule. On the basis of the Walsh diagram this state would be expected to have about the same bond angle as BH_2 in its ground state, since in both cases there is only one electron in the "bending" orbital $3a_1$. Actually, the $^3B_1$ state is observed to be the ground state and to have a bond angle very near to 180°—that is, is more properly described by $^3\Sigma_g^-$.

In the ground state of NH_2 there are three more electrons than in BeH_2; two of these must go into the $3a_1$ orbital and one into the $1b_1$ orbital. Therefore, on the basis of the Walsh diagram we expect a strongly bent molecule (similar to CH_2 in the $^1A_1$ state), in agreement with the observed angle of 103.4° [Dressler and Ramsay (38)].

Figure 75 shows graphically the observed structures of the

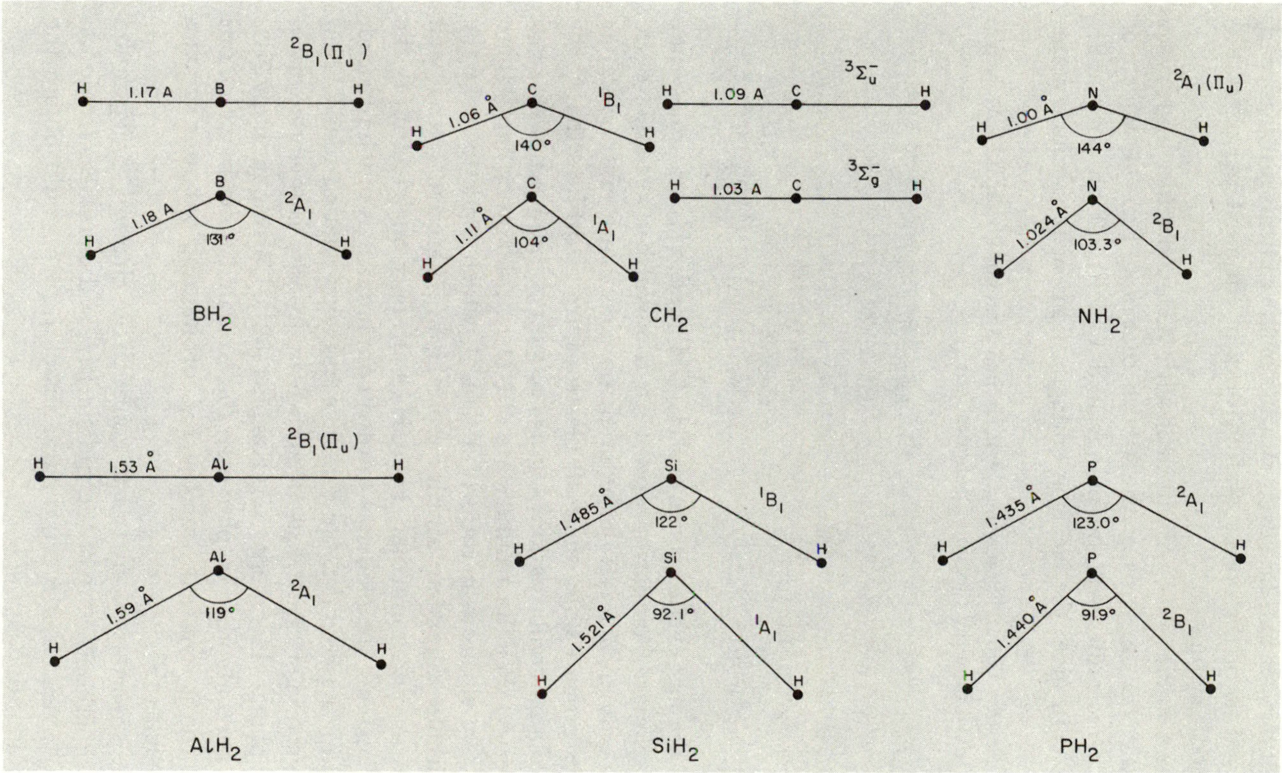

Fig. 75. Observed geometrical structures of dihydride radicals in their ground states and first excited states.

For CH_2 the structures for both singlet and triplet states are given; the actual ground state is the $^3\Sigma_g^-$ state. For SiH_2 so far only singlet states have been observed.

But, as we now [1988] know (see p. 16), the diagrams for the triplet CH_2 must be changed: $^3\Sigma_g^-$ to $^3B_1$ and $^3\Sigma_u^-$ to $^3A_2$, both with bond angles of 136°.

ground states and first excited states of the radicals just discussed as well as of the corresponding ones of the next period of the periodic system.

In a similar way we can deal with XH_3 molecules and try to predict on the basis of the Walsh diagram, Fig. 74, which of these molecules will be planar and which will be nonplanar. Table 18 gives the electron configurations of the ground states

Table 18. Ground States and First Excited States of Planar XH_3 Molecules as Derived from the Electron Configurations

Molecule	*Lowest Electron Configuration and Ground State*		*First Excited Electron Configuration and Resulting States*	
BH_3	$(1a_1')^2(2a_1')^2(1e')^4$	$^1A_1'$	$\ldots(2a_1')^2(1e')^3(1a_2'')$	$^3E''$, $^1E''$
CH_3	$(1a_1')^2(2a_1')^2(1e')^4(1a_2'')$	$^2A_2''$	$\left\{\begin{array}{l}\ldots(2a_1')^2(1e')^3(1a_2'')^2 \\ \ldots(2a_1')^2(1e')^4(3a_1')\end{array}\right.$	$^2E'$ $^2A_1'$
NH_3	$(1a_1')^2(2a_1')^2(1e')^4(1a_2'')^2$	$^1A_1'$	$\ldots(2a_1')^2(1e')^4(1a_2'')(3a_1')$	$^3A_2''$, $^1A_2''$

and the first excited states of BH_3, CH_3, and NH_3, assuming they are planar. Comparing now with the Walsh diagram, we see that only for BH_3 is it immediately obvious that it must be planar in its ground state, since all electrons are in orbitals that favor the planar conformation; but a spectrum of BH_3 has not yet been observed. For the ground state of NH_3, on the other hand, there are two electrons in an orbital ($1a_2''$), which tends to make the molecule nonplanar, and indeed NH_3 in its ground state is well known to be nonplanar. For CH_3, with only one electron in the $1a_2''$ orbital, the tendency toward nonplanarity is much reduced, and in fact considerable evidence has accumulated that CH_3 in its ground state, at least if zero-point energy is included, is in fact planar [Herzberg (57), Karplus (82)]. Correspondingly NH_3 has been found to be planar in those excited states in which one electron is removed from the $1a_2''$ orbital and only one remains [Douglas (33)].

It is perhaps of interest to consider the electron configuration of the ground state of the radical BH_4 and the isoelectronic ion CH_4^+. Assuming a tetrahedral structure, we would obtain the configuration

$$(1a_1)^2(2a_1)^2(1f_2)^5\ {}^2F_2.$$

However, on account of the Jahn-Teller effect (see p. 147f.) the resulting $^2F_2$ state must split into two states, $^2A_1$ and ^2E if the distorted conformation has $\boldsymbol{C}_{3v}$ symmetry. These same two states arise if we use from the beginning the point group $\boldsymbol{C}_{3v}$ for these radicals.[2] No spectra of either BH_4 or CH_4^+ have as yet been found in the laboratory in spite of much effort. Both radicals are of considerable chemical interest. In addition it has been suggested that CH_4^+ occurs in the interstellar medium, and indeed is responsible for the diffuse interstellar lines [Herzberg (58)(59)].

B. VIBRATIONAL LEVELS: VIBRONIC INTERACTIONS

(1) Nondegenerate electronic states

Just as for linear polyatomic molecules, the vibrational motion of a nonlinear molecule can be represented as a superposition of *normal vibrations*. As illustrations we present in Fig. 76 the normal

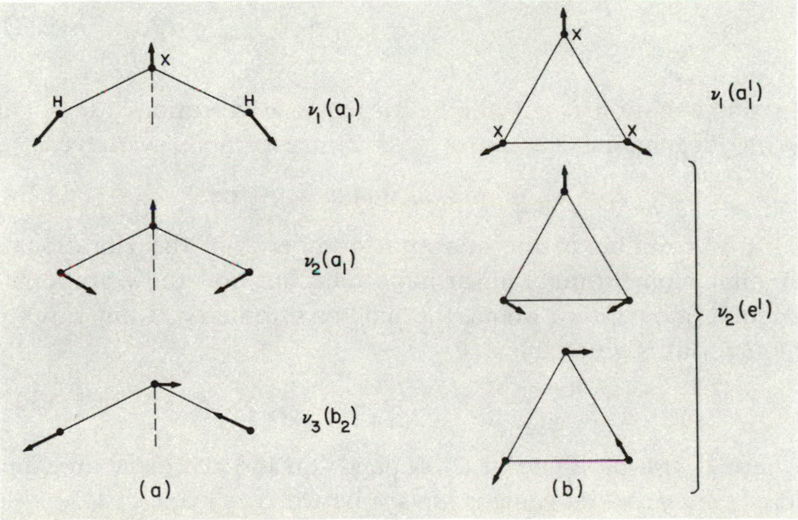

Fig. 76. Normal vibrations of (a) nonlinear XY_2 and (b) triangular X_3 molecules.

[2] Alternatively it may be that in the distorted conformation the point group is $\boldsymbol{D}_{2d}$. In that case the states $^2B_2$ and ^2E would result. Dixon (32a) has recently shown by *ab initio* calculations that this alternative in all probability applies to CH_4^+.

vibrations of a nonlinear XY_2 molecule (point group $\boldsymbol{C}_{2v}$) and a triangular X_3 molecule (point group $\boldsymbol{D}_{3h}$). The BH_2 radical discussed earlier is an example of the first type [Fig. 76(a)]; the H_3^+ ion is an example of the second type [Fig. 76(b)]. In the second type of molecule the analogues of the two vibrations ν_2 and ν_3 of the first type are degenerate with each other: they can be transformed into each other by a rotation of each displacement vector by 90° in the same sense and therefore have the same frequency $\nu_2(e')$ (similar to the bending vibration of a linear XY_2 molecule). Degenerate vibrations occur for all molecules with more-than-twofold axes of symmetry but do not occur for molecules without such symmetry axes except by accident.

The vibrational levels of a nonlinear polyatomic molecule in a nondegenerate electronic state are given by

$$G(v_1,v_2,v_3,\ldots) = \sum_i \omega_i\left(v_i + \frac{d_i}{2}\right) + \sum_{k\geq i}\sum_i x_{ik}\left(v_i + \frac{d_i}{2}\right)\left(v_k + \frac{d_k}{2}\right) + \ldots + \sum_{k\geq i}\sum_i g_{ik}l_il_k. \quad (128)$$

In this equation $l_i = 0$ for nondegenerate vibrations ($d_i = 1$). When degenerate vibrations ($d_i = 2$) are excited[3], we have

$$l_i = v_i,\ v_i - 2,\ v_i - 4,\ \ldots,\ 1 \text{ or } 0. \quad (129)$$

This l_i is similar to the quantum number l_i of the vibrational angular momentum of linear molecules, but now the vibrational angular momentum about the axis of symmetry is not simply $l_ih/2\pi$, but is given by

$$l_i\zeta_i\frac{h}{2\pi}, \quad (130)$$

where ζ_i is a parameter that depends on the geometry and the force constants. Its value is always between -1 and $+1$.

[3] As elsewhere in these lectures we shall restrict discussion to doubly degenerate vibrations. Triply degenerate vibrations occur only for molecules with cubic symmetry, and no spectra of radicals of such symmetry have yet been found.

As before, the vibrational eigenfunction is given in a first approximation by

$$\psi_v = \psi_{v_1}\psi_{v_2} \cdots \psi_{v_n}, \tag{131}$$

where the ψ_{v_i} are eigenfunctions corresponding to the different normal vibrations. Therefore the vibrational species (symmetry type) is obtained as the product of the species of the eigenfunctions ψ_{v_i}. For a totally symmetric vibration [for example, ν_1 in Fig. 76(a) or (b)], the ψ_{v_i} are totally symmetric for all values of v_i. However, for non-totally symmetric nondegenerate vibrations [e.g., ν_3 in Fig. 76(a)] the ψ_{v_i} are totally symmetric for even v_i and have the species of the normal vibration for odd v_i. Thus for a nonlinear XY_2 molecule the successive vibrational levels of the vibration $\nu_3(b_2)$ have the species

$$A_1,\ B_2,\ A_1,\ B_2,\ \ldots$$

for $v_3 = 0, 1, 2, 3, \ldots$, respectively, and similarly in other cases.

For degenerate vibrations the situation is somewhat more complicated (just as for linear molecules) in that for $v_i > 1$ several vibrational component levels arise. These are shown for an e' vibration of a $\boldsymbol{D}_{3h}$ molecule in Fig. 77. The higher levels of ν_2 in Fig. 76(b) would be represented by this figure. The same figure applies to degenerate vibrations of $\boldsymbol{C}_{3v}$ molecules if the primes attached to the species symbols are omitted. (For a more detailed discussion see *MM* II, p. 125f.)

As previously, the vibronic eigenfunction is in a first approximation given by

$$\psi_{ev} = \psi_e(q, 0)\psi_v(Q), \tag{132}$$

where $\psi_e(q, 0)$ is the electronic wave function for the equilibrium position ($Q = 0$). According to Eq. (132) the vibronic species is the direct product of the electronic species and the vibrational species. This direct product is formed in the same way as before. Thus in a B_1 electronic state of a $\boldsymbol{C}_{2v}$ molecule, if a B_2 vibration is singly excited [for example, ν_3 of Fig. 76(a)], the vibronic species is $B_1 \times B_2 = A_2$.

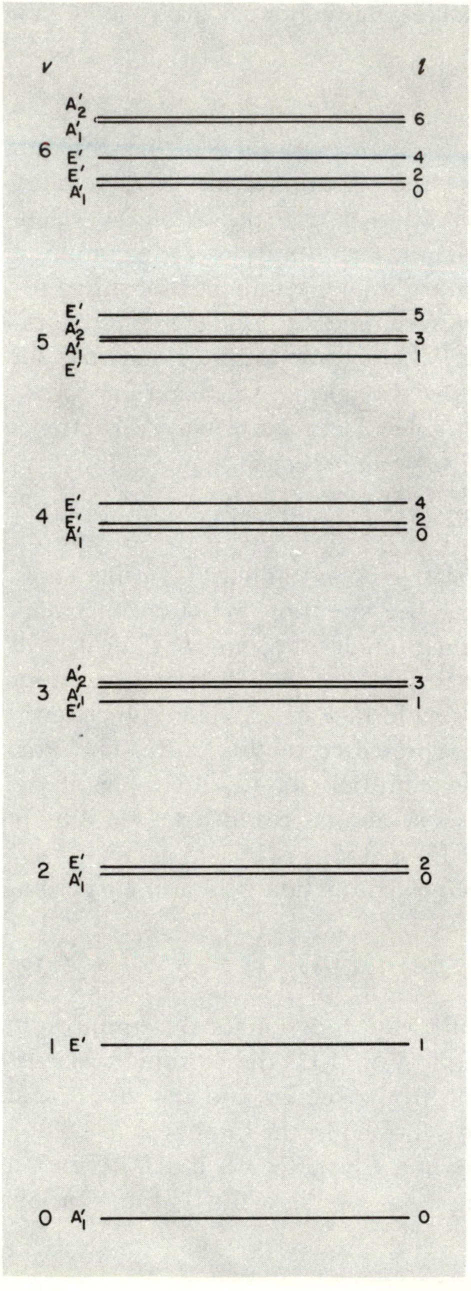

Fig. 77. Splitting of the higher vibrational levels of a degenerate (e') vibration of a $\boldsymbol{D}_{3h}$ molecule.

The same figure applies to other doubly degenerate vibrations of other point groups with appropriate changes of the species designations—for example, omitting all primes for point group $\boldsymbol{C}_{3v}$.

(2) Degenerate electronic states

Vibronic species. In a degenerate electronic state there are several sublevels of each vibrational level if degenerate vibrations are excited; for example, in a $\boldsymbol{D}_{3h}$ molecule in an E'' electronic state, if the degenerate vibration $\nu_2(e')$ [see Fig. 76(b)] is excited, we obtain the following species:

$$\begin{aligned} \text{for } v_2 &= 0\colon\ E'' \\ v_2 &= 1\colon\ A_1'' + A_2'' + E'' \\ v_2 &= 2\colon\ E'' + A_1'' + A_2'' + E'' \\ v_2 &= 3\colon\ A_1'' + A_2'' + E'' + 2E'' \end{aligned} \tag{133}$$

If vibronic interaction is taken into account, there will be as many component levels as there are vibronic species for each vibrational level. It must, however, be emphasized that degenerate vibronic levels cannot be further split by vibronic interactions; in particular, the lowest vibrational level always remains single with a degeneracy equal to that of the electronic state, irrespective of the strength of the vibronic interaction. Only the interaction with rotation (ro-vibronic interaction) can produce a splitting of this degeneracy. This ro-vibronic splitting is analogous to the Λ-type doubling of linear molecules.

Jahn-Teller theorem. In order to evaluate the magnitude of the vibronic splittings we must consider the splitting of the potential function for non-totally symmetric displacements of the nuclei. While for some displacements of this type the situation is the same as for linear molecules (see p. 99f.), there is according to Jahn and Teller (75) always at least one nontotally symmetric normal coordinate for which the splitting of the potential function is such that instead of two coinciding minima we have two separate minima at nonzero values of the particular normal coordinate. In other words, if the potential energy is plotted as a function of such a normal coordinate, the two resulting potential curves cross at the original equilibrium position with a nonzero angle. This situation is illustrated by Fig. 78.

Figure 78 gives the shape of the potential functions in only one projection. Actually, for a molecule with a threefold axis (for

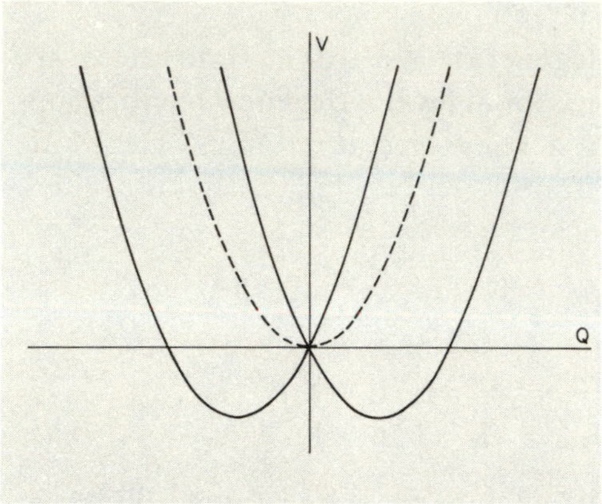

Fig. 78. Cross section through the potential function of a nonlinear molecule in a degenerate electronic state when vibronic interaction is large.

Q is a non-totally symmetric (usually degenerate) normal coordinate that gives rise to strong Jahn-Teller interaction. For comparison the potential function for zero vibronic interaction is shown as a broken-line curve.

example, CH_3I) the potential function in a plane perpendicular to the axis of symmetry will have three minima, as shown by the contour lines in Fig. 79. This figure, applied to CH_3I, shows

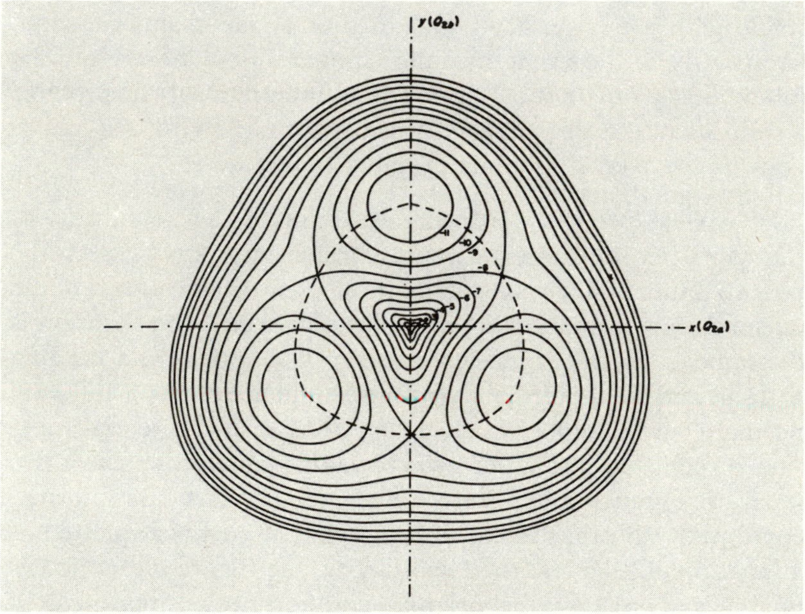

Fig. 79. Contour diagram of the lower part of the potential surface of a $\boldsymbol{C}_{3v}$ (or $\boldsymbol{D}_{3h}$) molecule in a degenerate electronic state (from *MM* III, p. 48).

Q_{2a} and Q_{2b} are two orthogonal components of a degenerate normal coordinate.

that the I atom in a degenerate electronic state will not have its equilibrium position on the axis of symmetry, but rather there will be three equivalent equilibrium positions somewhat away from the axis. The potential function as a whole still has C_{3v} symmetry. If the minima are deep—that is, if the energy required to bring the molecule from one minimum to the other is very high—the molecule may for most purposes be considered as an unsymmetrical one, of point group C_s. On the other hand if vibronic interaction is small, only a small amount of energy is required to go from one minimum to another—that is, for the motion of the molecule to cover the whole symmetrical potential surface. In that case, it is better to consider the molecule as symmetrical and study the modifications of the energy levels of such a symmetrical system by the vibronic interaction treated as a perturbation.

Vibronic energy levels. The splitting of the potential surface in a degenerate electronic state is called *static Jahn-Teller effect;* the splitting of the vibrational levels caused by this effect is called *dynamic Jahn-Teller effect*. In order to obtain these vibronic energy levels it is necessary to solve the wave equation when the potential function is of the type of Fig. 79. This has been done by a number of authors (see *MM* III, p. 45f). The result is that there is a splitting into as many different vibronic levels as there are species in the set (133) and corresponding ones for other cases. Frequently the simplifying assumption is made that the maxima between the minima along a circular path in Fig. 79 can be neglected, so that in effect there is a circular trough in which the motion of the molecule takes place. Under this simplifying assumption, the A_1 and A_2 vibronic levels that result according to (133) are not split, but the other vibronic splittings do occur.

For an X_3 molecule in an E' state, Child (17) has given for very small vibronic interaction the following formula for the vibronic levels corresponding to the degenerate vibration ν_2:

$$G(v_2, l) = \omega_2(v_2 + 1) \mp 2D\omega_2(l \pm 1). \qquad (134)$$

In this formula $D\omega_2$ is the depth of the "moat" below the peak at the origin (Fig. 79). According to Eq. (134), there are two levels for each l, which may be distinguished by a quantum number $j = l \pm \frac{1}{2}$. [For $l = 0$ only one level, corresponding to the upper signs in Eq. (134), arises.] Figure 80 shows the resulting levels obtained from this formula for $D = 0.04$.

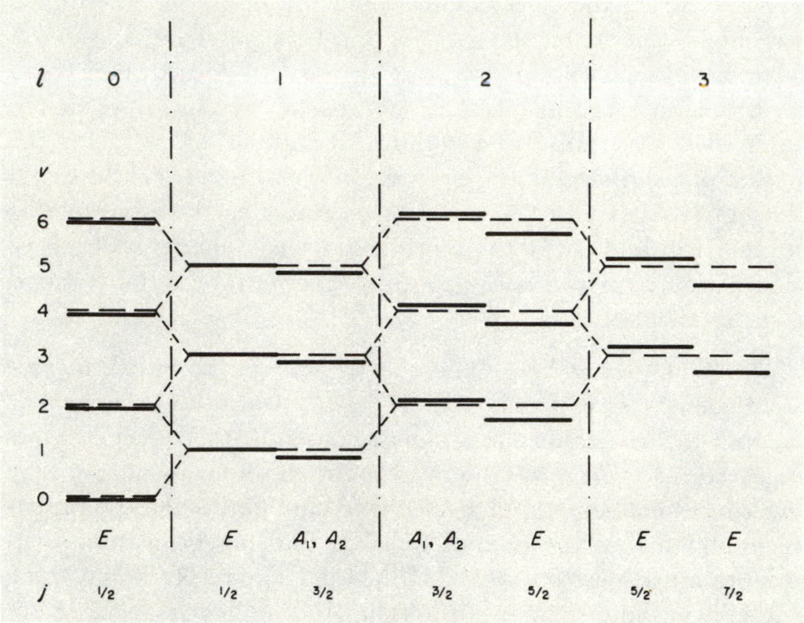

Fig. 80. Energy levels of a degenerate vibration in a degenerate electronic state of a $\boldsymbol{C}_{3v}$ (or $\boldsymbol{D}_{3h}$) molecule for very small vibronic interaction (from *MM* III, p. 55).

The vibronic levels are indicated by heavy solid lines. For comparison the levels obtained without introducing vibronic interaction are given as broken lines. The A_1, A_2 pairs are not split in this approximation.

The formula (134) holds only if $D < 0.05$. No explicit formulae have been given for larger D values, but Longuet-Higgins, Öpik, Pryce, and Sack (86) have numerically evaluated the energy levels for a number of D values. Figure 81 shows their results for $D = 2.5$. It is seen that a rather complicated vibronic structure arises: the groups of levels arising from different vibrational quantum numbers v (connected by sloping lines) overlap

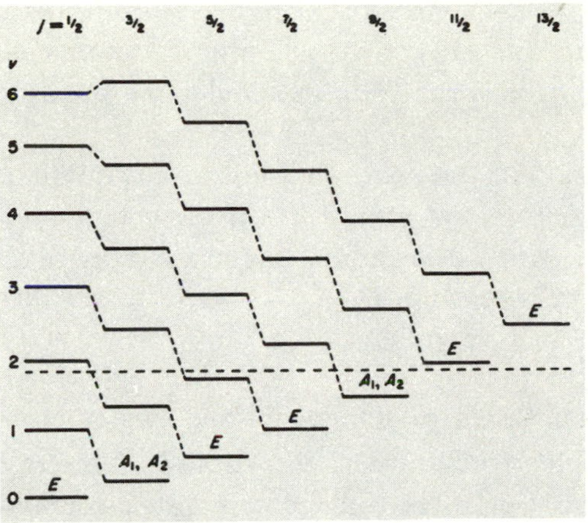

Fig. 81. Energy levels of a degenerate vibration in a degenerate electronic state of a $\boldsymbol{C_{3v}}$ (or $\boldsymbol{D_{3h}}$) molecule for large vibronic interaction (from *MM* III, p. 56).

Levels belonging to the same v value are connected by oblique broken lines. The horizontal broken line gives the energy of the potential minimum for zero vibronic interaction. The vibronic species is the same for all levels of a given column.

one another, while for zero vibronic interaction, in the harmonic approximation, they coalesce into single levels. Note, however, that degenerate vibronic levels, in particular the $v = 0$ level, are not split no matter how large the vibronic interaction.

C. ROTATIONAL LEVELS: RO-VIBRONIC INTERACTIONS

A nonlinear molecule, unlike a linear molecule, has no axis about which the moment of inertia vanishes. There are always three mutually perpendicular directions about which the moment of inertia has maximum or minimum values. These directions are called the *principal axes*. We designate the corresponding moments of inertia (called principal moments of inertia) by I_A, I_B, I_C, where I_A is the smallest, I_C the largest. If the molecule has an axis of symmetry, this axis is a principal axis; if the molecule has a plane of symmetry, there must be a principal axis perpendicular to that plane.

If two principal moments of inertia are equal, we have a *symmetric top*—a *prolate* symmetric top if $I_B = I_C$, and an *oblate* symmetric top if $I_A = I_B$. If all three principal moments of inertia are equal, we have a *spherical top*. The most general case ($I_A \neq I_B \neq I_C$) is that of the *asymmetric top*.

If a molecule has a more-than-twofold axis, it is necessarily a symmetric top. However, a less symmetrical molecule may be a symmetric top because accidentally two principal moments of inertia are equal. An example is the H_2S_2 molecule [see Winnewisser and Gordy (139)].

In the following discussion of the rotational levels of nonlinear molecules we shall consider only symmetric top molecules and asymmetric top molecules, since free-radical spectra of spherical top molecules have not yet been found.

(1) Symmetric top molecules

Nondegenerate vibrational levels in nondegenerate singlet electronic states. In singlet electronic states, the rotational-energy formula of the symmetric top including centrifugal-distortion terms is given by

$$F_v(J, K) = B_v J(J+1) + (A_v - B_v)K^2 - D_K K^4 - D_{JK} J(J+1)K^2 - D_J J^2(J+1)^2. \quad (135)$$

Here J, as usual, is the quantum number of the total angular momentum $\boldsymbol{J}$, while K is the quantum number of the component of $\boldsymbol{J}$ in the direction of the top axis. Figure 82 shows the

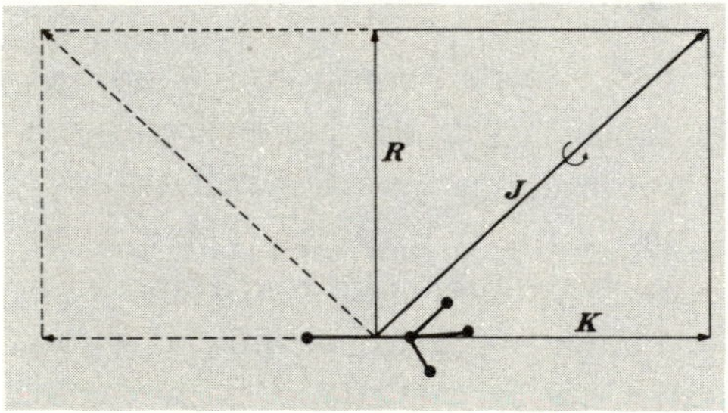

Fig. 82. Vector diagram for a symmetric top molecule. The broken-line vector diagram corresponds to a motion with exactly the same energy as the solid-line diagram.

corresponding vector diagram; while for linear and diatomic molecules the angular momentum about the top axis ($\boldsymbol{\Lambda}$) was produced only by the motion of the electrons, now it is produced by the motion of the heavy nuclei. Except for this difference and for the inclusion of the centrifugal-stretching terms, Eq. (135) is identical with Eq. (42) for the linear case.

The rotational constants B_v and A_v in Eq. (135) are given by the relations

$$B_v = B_e - \sum \alpha_i^B \left(v_i + \frac{d_i}{2}\right) + \cdots, \tag{136}$$

$$A_v = A_e - \sum \alpha_i^A \left(v_i + \frac{d_i}{2}\right) + \cdots, \tag{137}$$

where

$$B_e = \frac{h}{8\pi^2 c I_B{}^e}, \qquad A_e = \frac{h}{8\pi^2 c I_A{}^e} \tag{138}$$

are the equilibrium rotational constants and $I_A{}^e$ and $I_B{}^e$ the values of the moments of inertia for the equilibrium position. Equation (136) is the same as Eq. (106) used earlier for linear polyatomic molecules, while (137) is the analogous equation for the rotational constants A_v, which correspond to the moment of inertia about the axis of the top. The quantum number K can take the values of 0, 1, 2, . . . , while the quantum number J takes the values

$$J = K, K + 1, K + 2, \ldots . \tag{139}$$

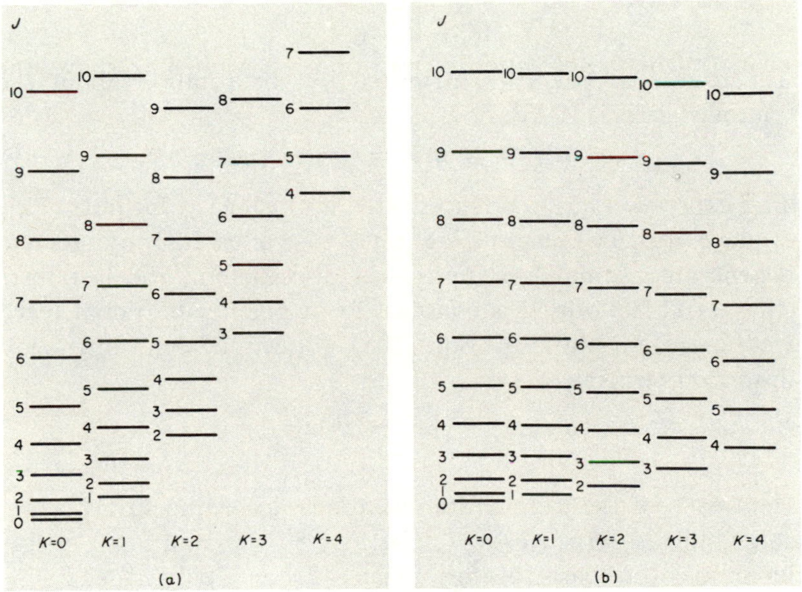

Fig. 83. Rotational energy levels of (a) a prolate and (b) an oblate symmetric top in a nondegenerate vibronic state (from *MM* III, p. 83).

The levels are arranged in different columns according to their K values.

The centrifugal-distortion terms $-D_K K^4 - D_{JK}J(J+1)K^2 - D_J J^2(J+1)^2$ in Eq. (135) are usually very small compared to the terms in B_v and A_v, and for the remainder of these lectures we shall disregard them. Figure 83 shows energy-level diagrams corresponding to Eq. (135) for a prolate and an oblate symmetric top. For the former $B = C$, as assumed in Eq. (135), while for the latter $A = B$, and C must be substituted for A in Eq. (135).

Degenerate vibrational levels in nondegenerate singlet electronic states. In degenerate vibrational levels a splitting arises with increasing rotation about the symmetry axis—that is, with increasing K—on account of the Coriolis forces that arise in a rotating molecule and act on the vibrational angular momentum. The angular momentum about the symmetry axis consists now of two components: the vibrational part, which is given by $\zeta_v(h/2\pi)$, and the rotational part, which is given by $K_r(h/2\pi)$. Thus we have

$$K = K_r \pm \zeta_v. \tag{140}$$

The rotational energy contributed by the rotation about the symmetry axis is $A_v K_r^2$, or

$$A_v(K \mp \zeta_v)^2 = A_v K^2 \mp 2A_v\zeta_v K + A_v\zeta_v^2. \tag{141}$$

This expression has to be substituted in the energy formula (135) in place of $A_v K^2$. Generalizing to the case of excitation of several degenerate normal vibrations and disregarding the last term $A_v\zeta_v^2$ in (141), which is constant for a given vibrational level, we obtain in place of Eq. (135) (omitting the centrifugal-distortion terms)

$$F_v(J, K) = B_v J(J+1) + (A_v - B_v)K^2 - 2A_v \sum(\pm l_i\zeta_i)K. \tag{142}$$

Here each of the degenerate vibrations according to Eq. (130) contributes a term $l_i\zeta_i$ to ζ_v. In Fig. 84 the rotational energy levels on the basis of formula (142) are plotted assuming $\sum(\pm l_i\zeta_i) = 0.4$. The difference from Fig. 83 is that all rotational levels with $K \neq 0$ are split in two, the magnitude of the splitting being constant for a given K (in the present approxi-

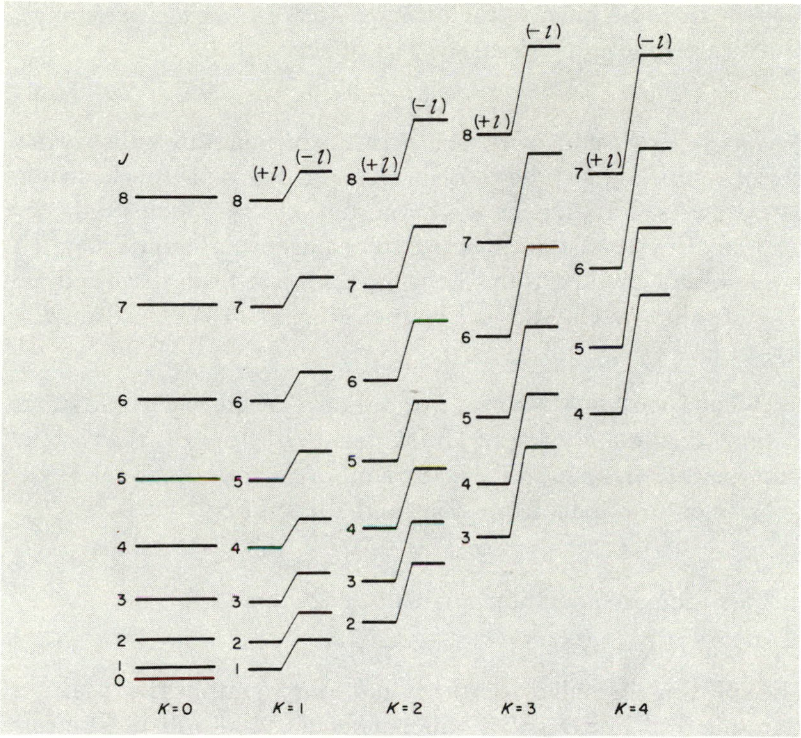

Fig. 84. Rotational energy levels of a prolate symmetric top in a degenerate vibronic state with $\zeta_v = 0.4$ (from *MM* III, p. 85).

In place of each single level with $K \neq 0$ in Fig. 83 there are now two levels marked $(+l)$ and $(-l)$ whose splitting increases linearly with K.

mation) and increasing linearly with K. The levels corresponding to the positive sign in Eq. (142) are called $(+l)$ levels, those with the negative sign $(-l)$ levels.

Degenerate electronic states. In a degenerate electronic state the electronic angular momentum about the symmetry axis has the value $\zeta_e(h/2\pi)$. In the rotating molecule this angular momentum interacts with $\boldsymbol{K}$, and a term similar to the one in Eq. (142) has to be added to Eq. (135). For $v = 0$ this term is

$$\mp 2A_v\zeta_e K. \tag{143}$$

If in the degenerate electronic state degenerate vibrations are excited, then we have to take the *total vibronic angular momentum*, which in a first approximation can be written

$$\zeta_t = \zeta_e \pm \zeta_v. \tag{144}$$

We have therefore in every degenerate vibronic state a first-order Coriolis splitting, but the cause of this splitting is partly electronic and partly vibrational in origin. Equation (144) holds only for very small vibronic interaction. For larger interaction much more complicated relations hold, which have been discussed for X_3 molecules by Child and Longuet-Higgins (18) (see also *MM* III, p. 64).

Multiplet electronic states. For nondegenerate electronic states in general, the analogue of Hund's case (b) applies; that is, the total angular momentum $\boldsymbol{J}$ is the sum of $\boldsymbol{N}$, the rotational angular momentum apart from spin, and the spin $\boldsymbol{S}$; that is,

$$\boldsymbol{J} = \boldsymbol{N} + \boldsymbol{S};$$

and for the corresponding quantum numbers

$$J = N + S, N + S - 1, \ldots, |N - S|. \tag{145}$$

The splitting formulae are somewhat more complicated than for diatomic and linear polyatomic molecules and will not be discussed here (see *MM* III, pp. 88 and 90). For degenerate electronic states no detailed discussion of multiplet splitting has yet been given.

Symmetry properties of rotational levels. As in the case of diatomic and linear polyatomic molecules, we may distinguish *positive* (+) and *negative* (−) rotational levels, depending on whether the overall eigenfunction remains unchanged or changes sign for a reflection at the origin. However, for nonplanar molecules such an inversion gives a different conformation, and therefore both the sum and the difference of the eigenfunctions corresponding to the two conformations are solutions of the wave equation: we have a double degeneracy, one of the levels being "positive," the other "negative." Only if the potential barrier between the two conformations is small (as in NH_3) does a splitting of this degeneracy arise, and then the *parity* (+ or −)

is of importance. For planar molecules the rotational levels are either + or −, but this distinction is not important, since there are usually other symmetry properties that are equivalent to + or −.

The ro-vibronic wave function ψ_{evr} must belong to one of the species of the point group of the molecule; that is, for a $\boldsymbol{C}_{3v}$ molecule the ro-vibronic species can be only A_1 or A_2 or E; similarly, for a $\boldsymbol{D}_{3h}$ molecule it can be only A_1', A_2', A_1'', A_2'', E', or E''. In Fig. 85 these symmetry properties are shown for the rotational levels of a $\boldsymbol{D}_{3h}$ molecule in an A_1' and an E' vibronic state. If the spin of the identical nuclei is zero, only A_1' rotational levels can appear (A_1 for $\boldsymbol{C}_{3v}$); all the others are missing. This is analogous to the fact that for a homonuclear diatomic molecule with $I = 0$ only the s levels are present. For $I = \frac{1}{2}$ of the identical nuclei, levels of both species A_2 and E can appear in the ratio 2:1. This holds independently of whether in $\boldsymbol{D}_{3h}$ a prime or double prime is attached to the species symbol. For $I = 1$, all three types of rotational levels can appear in the ratio 10:1:8 for A_1, A_2, and E. Thus, there is a very marked alternation of statistical weights of the rotational levels, which is characteristically different from diatomic and linear polyatomic molecules.

l-type doubling. As can be seen from Fig. 85, there are, for $K = 3$, 6, 9, . . . of a totally symmetric vibronic state (A_1'), and similarly for the ($+l$) levels of $K = 1$, 4, 7, . . . and the ($-l$) levels of $K = 2$, 5, 8, . . . of a degenerate state (E'), always two rotational levels for a given J value, one of species A_1, the other of species A_2. In principle there is always a splitting of these pairs of levels, but this splitting has an easily observable magnitude only for the $K = 1$ levels of an E vibronic level [see Fig. 85(b)]. For this case the splitting is given by

$$\Delta\nu = q_v J(J + 1). \tag{146}$$

This doubling is quite analogous to the l-type doubling in linear polyatomic molecules; in both cases it is the coupling of the vibrational angular momentum ($\boldsymbol{l}$) with the angular momentum of overall rotation ($\boldsymbol{J}$) that causes the splitting.

In a degenerate electronic state a similar doubling arises even without excitation of degenerate vibrations. It corresponds to the electronic angular momentum (ζ_e) and is called *j-type doubling*.

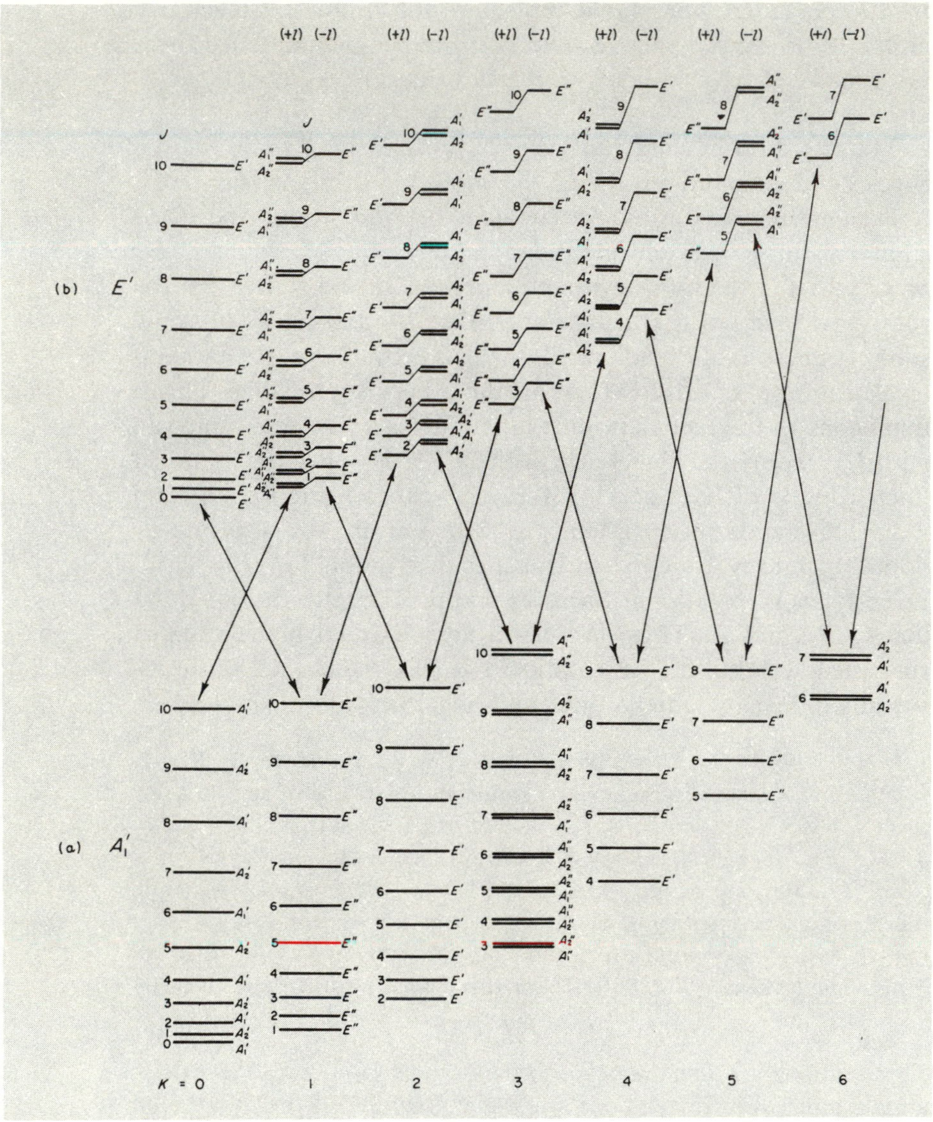

Fig. 85. Rotational energy levels in A_1' and E' vibronic states of $\boldsymbol{D}_{3h}$ molecules with ro-vibronic species designations (from *MM* III, p. 91).

The oblique arrows give the transitions for a $\perp$ band (see p. 180). By dropping all single and double primes we obtain the corresponding diagram for a $\boldsymbol{C}_{3v}$ or $\boldsymbol{D}_3$ molecule.

(2) Asymmetric top molecules

General rotational-energy formulae. For an asymmetric top ($I_A \neq I_B \neq I_C$) the rotational energy levels, neglecting centrifugal-distortion corrections, can be represented by

$$F_v(J_\tau) = \tfrac{1}{2}(B_v + C_v)J(J+1) + [A_v - \tfrac{1}{2}(B_v + C_v)]W_{J_\tau}{}^v. \quad (147)$$

In this formula A_v, B_v, and C_v are the rotational constants corresponding to the moments of inertia I_A, I_B, I_C in the vibrational level v. As in Eqs. (136) and (137) we have

$$\begin{aligned} A_v &= A_e - \sum \alpha_i{}^A(v_i + \tfrac{1}{2}) + \cdots, \\ B_v &= B_e - \sum \alpha_i{}^B(v_i + \tfrac{1}{2}) + \cdots, \\ C_v &= C_e - \sum \alpha_i{}^C(v_i + \tfrac{1}{2}) + \cdots, \end{aligned} \quad (148)$$

$$A_e = \frac{h}{8\pi^2 c I_A{}^e}, \qquad B_e = \frac{h}{8\pi^2 c I_B{}^e}, \qquad C_e = \frac{h}{8\pi^2 c I_C{}^e}.$$

$W_{J_\tau}{}^v$ is a quantity corresponding to K^2 of the symmetric top; it takes $2J + 1$ different values for each value of J. Thus we have $2J + 1$ sublevels for each J rather than $J + 1$, as in the case of the symmetric top; in other words, each level with a given J and K of the symmetric top is split into two, except the one with $K = 0$. The magnitude of the quantity $W_{J_\tau}{}^v$ is obtained from somewhat complicated algebraic equations of order J, which depend on the *asymmetry parameter*

$$b = \frac{C_v - B_v}{2[A_v - \tfrac{1}{2}(B_v + C_v)]}. \quad (149)$$

It may be noted that $b = 0$ for a prolate symmetric top ($B = C$) and $b = -1$ for an oblate symmetric top ($A = B$).

In order to distinguish the $2J + 1$ sublevels for a given J, the parameter

$$\tau = J, J-1, J-2, \ldots, -J \quad (150)$$

is assigned to the levels in the order of their energy; that is, J_{+J} designates the highest, J_{+J-1} the second highest, . . . , J_{-J} the lowest of the levels.

An alternative form of the energy formula is

$$F_v(J_\tau) = \tfrac{1}{2}(A_v + C_v)J(J+1) + \tfrac{1}{2}(A_v - C_v)E_{J_\tau}{}^v, \quad (151)$$

where $E_{J_\tau}{}^v$ can also be expressed in terms of b but is usually expressed in terms of the parameter

$$\kappa = \frac{2[B_v - \frac{1}{2}(A_v + C_v)]}{A_v - C_v} = -\frac{1 + 3b}{1 - b}. \tag{152}$$

It takes the value -1 for the prolate and $+1$ for the oblate symmetric top.

Instead of using the parameter τ, many authors use a double subscript, the first one giving the K value of the corresponding level in the prolate symmetric top and the second one that of the corresponding level in the oblate symmetric top. These K values are called here K_a and K_c, although many authors prefer to call them K_{-1} and K_{+1} (-1 and $+1$ being the values of κ for the prolate and oblate symmetric top, respectively). Thus one has the designation $J_{K_aK_c}$ or $J_{K_{-1}K_{+1}}$. In Fig. 86 the energy levels of an asymmetric top are shown schematically with the two alternative notations added.

If the asymmetry parameter b is very small, the asymmetric top formula, Eq. (147), approaches the symmetric top formula, Eq. (135). This is because for small b the quantity $W_{J_\tau}{}^v$ approaches K^2. If this is substituted in Eq. (147) and the resulting equation compared with Eq. (135), we see that the effective B value for a slightly asymmetric top is $\frac{1}{2}(B_v + C_v)$. However, even for fairly small asymmetry there will be a splitting (*asymmetry doubling*) of all levels with $K \neq 0$. This splitting is largest for $K = 1$, as shown in more detail in *MM* III, p. 107. One finds for the effective B values of the two sublevels of $K = 1$ the expression

$$B^1{}_{\text{eff}}{}^d = \tfrac{1}{4}(B_v + 3C_v), \qquad B^1{}_{\text{eff}}{}^c = \tfrac{1}{4}(3B_v + C_v), \tag{153}$$

and therefore for the magnitude of the asymmetry doubling for $K = 1$

$$\Delta\nu_{cd}{}^{K=1} = \tfrac{1}{2}(B_v - C_v)J(J + 1); \tag{153a}$$

that is, the doubling is proportional to the difference $B_v - C_v$ (which vanishes for the symmetric top) and increases with $J(J + 1)$.

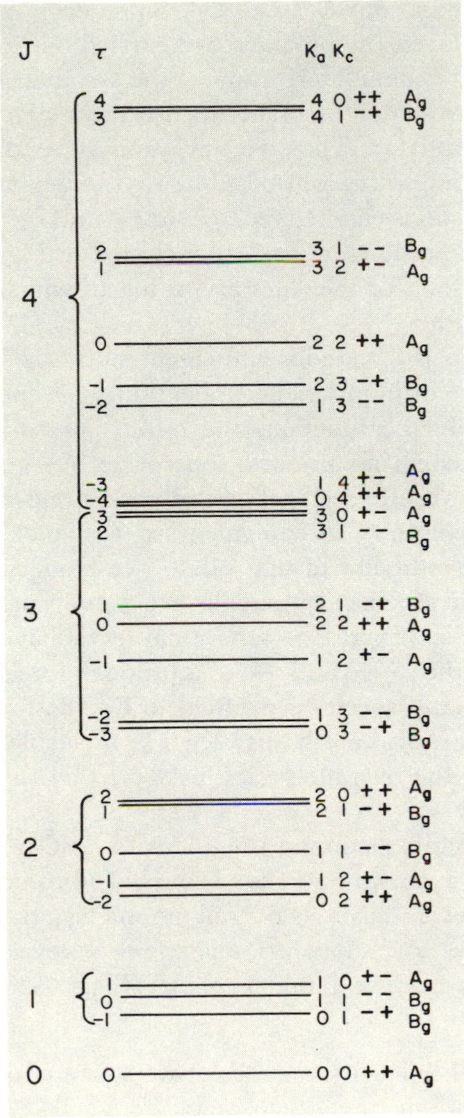

Fig. 86. Rotational energy levels of an asymmetric top molecule with $\kappa = -0.2$.

The τ values are given at left, the K_aK_c values at right. The asymmetric rotor species $++$, $+-$, $-+$, $--$ and the overall species for the case of a totally symmetric state of a C_{2h} molecule are also indicated.

Symmetry properties of rotational levels. The eigenfunctions ψ_r of the asymmetric rotor are symmetric or antisymmetric with respect to rotations by 180° about any one of the principal axes *a*, *b*, or *c*. We call these operations C_2^a, C_2^b, C_2^c. Since each of these operations can be replaced by the other two carried out in succession, it is sufficient for a characterization of the symmetry properties of the asymmetric rotor functions to use their symmetry with respect to C_2^c and C_2^a. Thus we have four types of rotational levels corresponding to the four possible ways in which the corresponding eigenfunctions can behave toward C_2^c and C_2^a: $++$, $+-$, $-+$, $--$. These designations are included in Fig. 86. We refer to *MM* II, p. 50, for the rules according to which this assignment must be made.

If the molecule has symmetry, the vibronic eigenfunction ψ_{ev} has symmetry; therefore we obtain, by multiplication with the symmetry of the rotational eigenfunction, the *overall symmetry properties*. The question is which asymmetric top species ($++$, $+-$, . . .) go with which overall species. For this correlation a reflection at a plane of symmetry is equivalent to a twofold rotation about an axis perpendicular to that plane [see Hougen (74)]. With this in mind we see readily that for a $\boldsymbol{C}_{2h}$ molecule in an A_g vibronic level, the $++$ and $+-$ rotational levels have A_g overall symmetry and the $-+$ and $--$ rotational levels have B_g symmetry. The overall species are added to Fig. 86 for this case. For other types of vibronic levels of a $\boldsymbol{C}_{2h}$ molecule, one has simply to multiply the overall species just given by the species of the vibronic level.

Similarly, for a $\boldsymbol{C}_{2v}$ molecule, assuming the *a* axis to be $C_2(z)$ and the *c* axis to be the *x* axis, we find that in an A_1 vibronic state the $++$ rotational levels have A_1 overall symmetry, the $+-$ levels have B_2, the $-+$ levels have A_2, and the $--$ levels have B_1. Further examples may be found in *MM* II, p. 462f, and *MM* III, p. 109f.

In addition to these overall symmetries, we also have to consider the *parity* ($+$ or $-$). As for symmetric top molecules, for nonplanar asymmetric top molecules there are two levels for each rotational level, of which one is $+$ and the other is $-$. These two levels are appreciably split only when the potential barrier preventing an inversion is very small. For planar asymmetric top molecules the parity for totally symmetric vibronic levels can be obtained simply from the behaviour with respect to C_2^c, while it is opposite to this behaviour for vibronic levels whose eigenfunctions are antisymmetric with respect to the plane of the molecule. The consideration of parity is, however, not important for asymmetric top molecules if the full symmetry of the point group is used for the overall species.

Spin splitting. Since an asymmetric top molecule cannot have an electronic orbital angular momentum, spin-orbit interaction is in general small, similar to Hund's case (b) in linear molecules. For $S = \frac{1}{2}$—that is, doublet states—the two component levels can be represented by

$$\begin{aligned} F_1(N_\tau) &= F_0(N_\tau) + \tfrac{1}{2}\gamma N_\tau, \\ F_2(N_\tau) &= F_0(N_\tau) - \tfrac{1}{2}\gamma(N_\tau + 1), \end{aligned} \tag{154}$$

where N is the quantum number of the total angular momentum apart from spin, corresponding to J, and where τ [as in Eq. (150)] distinguishes levels of the same N; $F_0(N_\tau)$ is identical with $F_v(J_\tau)$ of Eq. (147) except that J_τ is everywhere replaced by N_τ.

According to Raynes (119) the splitting constant γ depends on both K and N as follows:

$$\gamma = \kappa \frac{K^2}{N(N+1)} + \mu \pm \tfrac{1}{2}\eta_K, \tag{155}$$

where κ, μ, and η_K are constants. The last term corresponds to a difference in spin splitting for the two components of an asymmetry doublet. It is appreciable only for $K = 1$.

For triplet states of asymmetric top molecules an additional term arises in the splitting formula similar to that for $^3\Sigma$ states of diatomic molecules. More details may be found in *MM* III, pp. 90 and 118.

(3) Quasi-linear molecules

When a molecule is only slightly bent, with increasing amplitude of the bending vibration a gradual transition of the rotational levels takes place from those of a nearly symmetric top to those of a linear molecule. Such a system is called a quasi-linear molecule.

The potential energy of such a molecule as a function of the displacement x from the linear conformation is represented by a curve of the type shown in Fig. 87: it has a potential hump at $x = 0$. A variety of mathematical expressions have been used to represent such a potential function. Thorson and Nakagawa (131) used

$$V = \tfrac{1}{2}kx^2 + \frac{K_B}{c^2 + x^2}, \tag{156a}$$

while Dixon (32) chose

$$V = \tfrac{1}{2}kx^2 + \alpha e^{-\beta x^2}. \tag{156b}$$

The potential surface, of which Fig. 87 presents a cross-section, is symmetric about the axis corresponding to the linear conformation. Thus, the two minima shown in Fig. 87 are not really separate: we can go from one to the other by a simple rotation of the molecule.

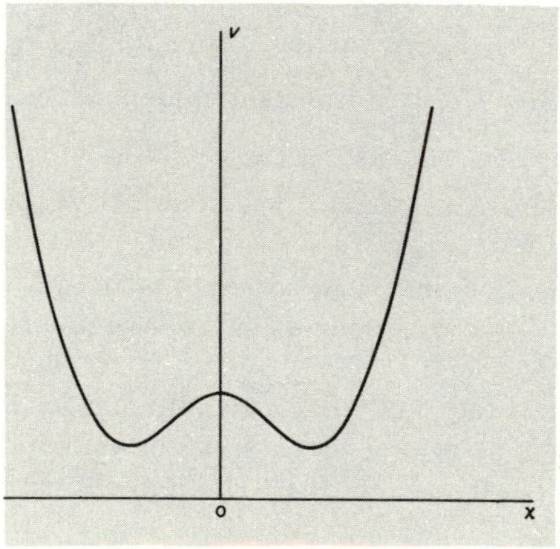

Fig. 87. Potential energy as a function of the bending coordinate (x) in a quasi-linear molecule (from *MM* III, p. 120).

A slightly bent molecule, while strictly speaking an asymmetric top, is always fairly close to being a prolate symmetric top, and therefore the quantum number K is fairly well defined. As the vibrational energy increases or the potential hump decreases, K goes over into l, the quantum number of the vibrational angular momentum of the linear molecule. Figure 88 shows the correlation of the levels between the linear and bent conformations. In this figure the barrier is assumed to increase from left to right.

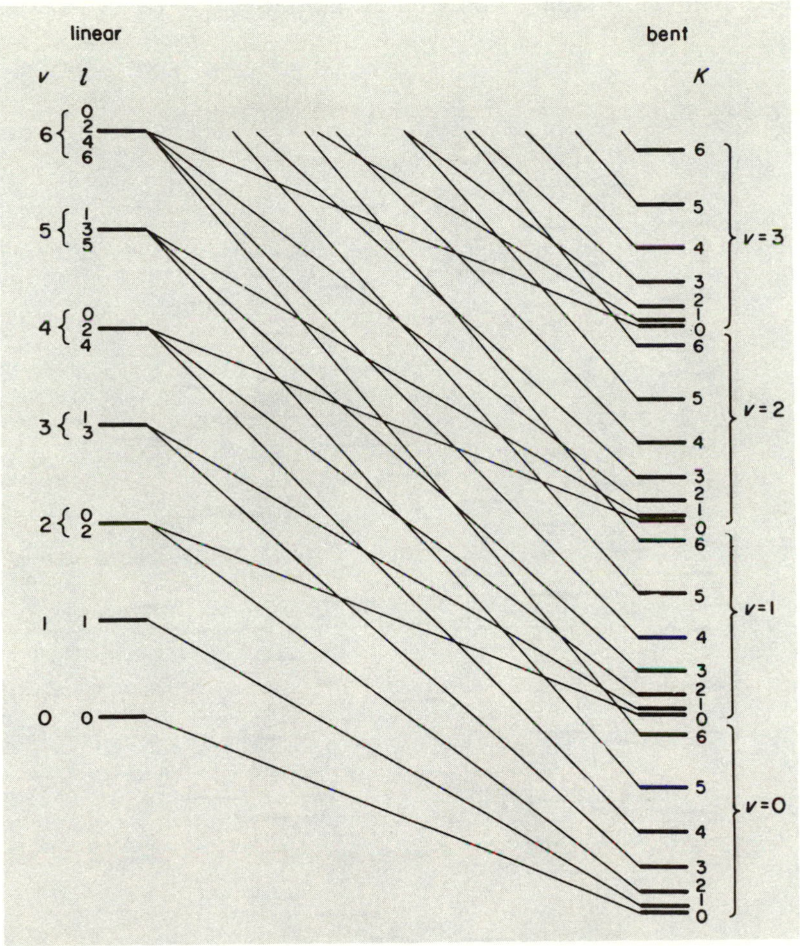

Fig. 88. Correlation of energy levels of linear and bent molecules in a non-degenerate electronic state (from *MM* III, p. 121).

The height of the barrier increases from left to right; the energy curves are only qualitatively correct.

For a given height of the barrier Fig. 89 shows the energy levels as a function of K. It is seen that below the barrier the energy levels for a given value of v increase quadratically with K. This is no longer the case near or above the barrier, since in the linear case for each v alternately only even and odd l, that is—K values exist.

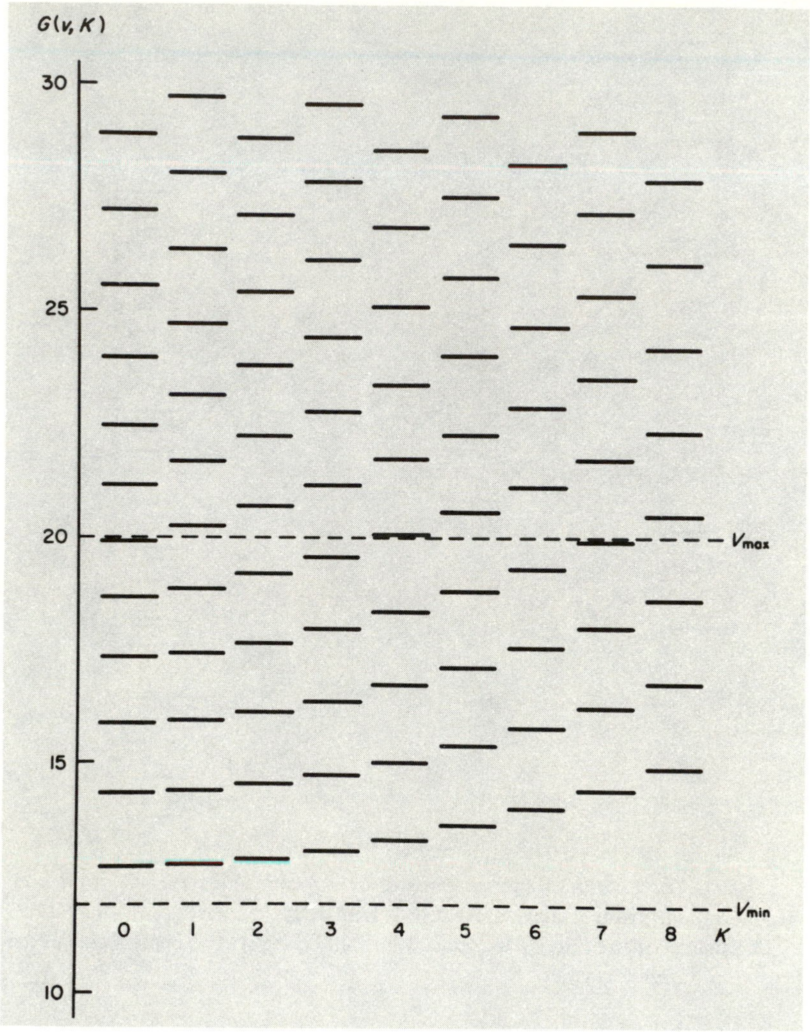

Fig. 89. Energy levels of the bending vibration of a quasi-linear molecule near the top of the barrier, after Dixon (32) (from *MM* III, p. 123).

The two broken lines give the energies of the potential minima and of the maximum in Fig. 87. $G = 0$ corresponds here to $\alpha = 0$ in Eq. (156b), not to the energy of the minimum.

As a result of the transition from the bent to the linear conformation, the rotational constant A_v increases much more than linearly with v, and the same is true for the centrifugal-stretching constant D_K, which can assume quite large values. It may be seen from Fig. 89, as first pointed out by Dixon (32), that the vibrational differences, which below the barrier decrease regularly, reach a minimum near the barrier and then increase.

D. TRANSITIONS, EXAMPLES

(1) Rotation and rotation-vibration spectra

The rotation spectra of nonlinear polyatomic radicals are, of course, entirely similar to those of stable molecules. They can be observed in the microwave region [see the monographs by Townes and Schawlow (132), Gordy, Smith, and Trambarulo (47), Sugden and Kenney (127), and Wollrab (141)] or in the far infrared, if the molecule has a permanent dipole moment. For molecules without a dipole moment pure rotational transitions can be observed only in the Raman effect [see *MM* II, chap. I, and the more recent review of Stoicheff (124)]. While no Raman spectra of free radicals have as yet been obtained, the microwave spectra of two fairly long-lived radicals have been studied: CF_2 by Powell and Lide (114) and SiF_2 by Rao and

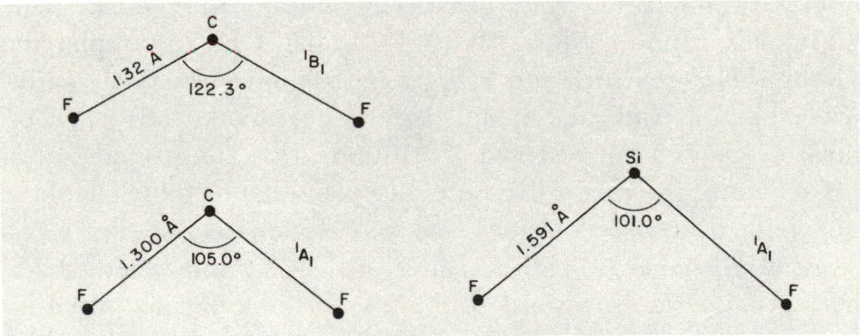

Fig. 90. Observed geometrical structures of CF_2 and SiF_2.

The data for the ground states are from microwave spectra (see text), the data for the excited state of CF_2 from the analysis of the ultraviolet spectrum by Mathews (89). It is not entirely certain whether the observed $^1A_1$ state or the as yet unobserved lowest triplet state ($^3B_2$) is the ground state.

Curl (117). All observed transitions correspond to the selection rules

$$\Delta J = 0, \pm 1; \qquad ++ \leftrightarrow --; \qquad +- \leftrightarrow -+, \quad (157)$$

which are expected to hold (see *MM* II, p. 55) for an asymmetric top when the dipole moment lies in the principal axis of intermediate moment of inertia (*b* axis). In addition the approximate rule

$$\Delta K_a = \pm 1 \quad (157a)$$

applicable to slightly asymmetric prolate tops is found to be obeyed (see *MM* III, p. 245 and p. 248). From the observed lines the rotational constants A_0, B_0, C_0, and from them the geometrical structures, have been determined as illustrated in Fig. 90. For CF_2 there is good agreement with the structure obtained from the electronic spectrum (see below).

Infrared vibration spectra of a large number of free radicals have been observed in solid matrices at low temperature, but as yet few radicals have been studied in the gaseous phase, none at high resolution. Probably the first was the infrared spectrum of CF_2 observed by Herr and Pimentel (52). They observed a single band at 1110 cm^{-1}, probably ν_3. Carlson and Pimentel (13) have observed the infrared spectrum of CF_3, finding three of the four fundamentals and showing that CF_3, unlike CH_3, is nonplanar. Khanna, Hauge, Curl, and Margrave (83) have observed ν_1 and ν_3 of the SiF_2 radical. In no case has rotational structure of infrared bands been resolved. For nondegenerate ground states this structure is expected to be entirely similar to those of infrared bands of stable molecules (see *MM* II, chap. IV). For electronically degenerate ground states, just as for linear molecules in Π ground states, several subbands are expected for some of the infrared bands on account of vibronic interactions—namely, when in the upper state one or more degenerate vibrations are excited. Since no such case has been observed for any free radical, we shall not discuss the expected band structure.

(2) Electronic transitions

Allowed electronic transitions. As for linear polyatomic molecules, an electronic transition is allowed if

$$R_{e'e''} = \int \psi_e'^* M \psi_e'' \, d\tau_e \tag{158}$$

is different from 0, or, in other words, if the product $\psi_{e'}\psi_{e''}$ has the same species as the dipole moment M. If the molecule has a different symmetry in the upper and lower state, we must use the common point group of lower symmetry in order to decide whether a given electronic transition is allowed.

In addition to the symmetry selection rule based on (158), we must also consider the spin. Just as for atoms and diatomic molecules, we have the selection rule

$$\Delta S = 0, \tag{159}$$

which holds as long as spin-orbit coupling is weak.

As an example, let us consider the allowed electronic transitions for a $\boldsymbol{C}_{2v}$ molecule or a molecule in which the common elements of symmetry in the upper and lower state belong to point group $\boldsymbol{C}_{2v}$. On the basis of Table 13(a), by multiplying the symmetry types and remembering that the components M_x, M_y, and M_z of the dipole moment have the species B_1, B_2, and A_1, respectively, we see that A_1–B_1 and A_2–B_2 transitions are allowed with the dipole moment in the x direction, A_1–B_2 and A_2–B_1 transitions are allowed if the dipole moment lies in the y direction, and A_1–A_1, A_2–A_2, B_1–B_1, B_2–B_2 are allowed if the dipole moment lies in the z direction. The only electronic transitions of a $\boldsymbol{C}_{2v}$ molecule for which no component of the dipole moment leads to a nonzero value of the integral in Eq. (158) are A_1–A_2 and B_1–B_2. They are forbidden electronic transitions. We indicate the forbidden nature of such transitions by writing $A_1 \not\leftrightarrow A_2$, $B_1 \not\leftrightarrow B_2$.

The selection rules for other point groups are readily obtained in a similar manner. Table 19 gives the results for the point group $\boldsymbol{D}_{3h}$ to which, for example, the CH_3 radical belongs. For other point groups see Table 9 of *MM* III, p. 132.

Table 19. Transition Moments of Electronic Transitions of Molecules Belonging to Point Group $\boldsymbol{D}_{3h}$

$\boldsymbol{D}_{3h}$	A_1'	A_2'	A_1''	A_2''	E'	E''	
	f.	*f.*	*f.*	M_z	$M_{x,y}$	*f.*	A_1'
		f.	M_z	*f.*	$M_{x,y}$	*f.*	A_2'
			f.	*f.*	*f.*	$M_{x,y}$	A_1''
				f.	*f.*	$M_{x,y}$	A_2''
					$M_{x,y}$	M_z	E'
						$M_{x,y}$	E''

The species of one state participating in the transition is given in the top row of the table, that of the other state in the column at the right. M_x, M_y, M_z give the orientation of the dipole moment for the particular transition. $M_{x,y}$ indicates that M_x and M_y are equivalent; *f.* refers to a forbidden transition. The table applies to point group $\boldsymbol{C}_{3h}$ if the subscripts 1 and 2 are omitted.

Forbidden electronic transitions. Forbidden electronic transitions can occur, as for linear polyatomic molecules, because of

(1) magnetic dipole and electric quadrupole radiation,
(2) spin-orbit interaction, and
(3) vibronic interactions.

The first two types of forbidden transitions are the same as for diatomic molecules and will not be discussed here (see *MM* III, p. 134f). The third is the same as for linear polyatomic molecules, but it needs some amplification.

As we have seen earlier, the vibronic transition moments $R_{e'e''v'v''}$ may be different from zero even if $R_{e'e''} = 0$—that is, even if the transition is electronically forbidden. However, $R_{e'e''v'v''}$ can be different from zero only for certain vibrational transitions, which, as for linear molecules, follow the opposite selection rule to those of allowed transitions (namely, those that make the integrand $\psi_e'^*\psi_v'^*M\,\psi_e''\psi_v''$ totally symmetric).

As an example, let us consider the CH_3 free radical. On the basis of the electron configuration (see Table 18) the first excited state is expected to be an E' state. According to Table 19 this state cannot combine with the A_2'' ground state, assuming the molecule to be planar in both states. However, this forbidden transition may yet occur (even though weakly) on account of vibronic interaction, but only if the vibrational quantum number v_k of an antisymmetric vibration changes by an odd number ($\Delta v_k = 1, 3, \ldots$) rather than an even number as for allowed electronic transitions (see below). The same transition may occur as an allowed transition if the radical, as seems likely, is not planar in the excited state, since then the less restrictive selection rules for $\boldsymbol{C}_{3v}$ apply. But this transition has not yet been observed. A similar forbidden transition (E''–A_1') connects the first excited state of BH_3 with its ground state but has also not yet been observed.

(3) Vibrational structure of electronic transitions

Transitions between nondegenerate electronic states. For nonlinear molecules the same considerations concerning vibrational transitions between nondegenerate electronic states hold as for linear molecules; that is,

$$\psi_v'^{*}\psi_v'' \textit{ must be totally symmetric.} \tag{122}$$

In other words, only vibrational levels of the same vibrational species can combine with one another. In this way, for totally symmetric vibrations we get the same kind of progressions and sequences as for linear molecules (see p. 108f.). For non-totally symmetric vibrations we have again the selection rule

$$\Delta v_k = 0, \pm 2, \pm 4, \ldots \tag{160}$$

combined with

$$\Delta l_k = 0. \tag{160a}$$

Again, as for linear molecules, among the transitions allowed by Eqs. (160) and (160a) those with $\Delta v_k = 0$ are usually much more intense than those with $\Delta v_k = \pm 2, \pm 4, \ldots$.

Transitions between electronic states at least one of which is degenerate. The vibrational structure of electronic transitions involving degenerate electronic states, such as E–A or E–E transitions, is complicated by the occurrence of the Jahn-Teller effect when degenerate vibrations are involved. As an example, Fig. 91 shows an energy-level diagram giving the vibrational

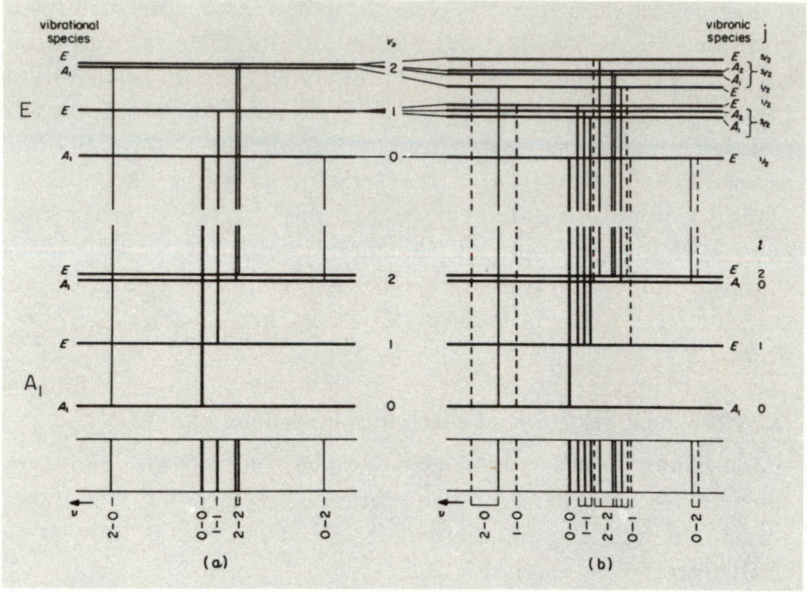

Fig. 91. Vibrational transitions in a degenerate vibration ν_k for a ^1E–$^1A_1$ electronic transition of a C_{3v} molecule (a) without and (b) with vibronic (Jahn-Teller) splitting (from *MM* III, p. 161).

Compare the caption of Fig. 66. The quantum number j at the right is well defined only if the three identical potential minima in the excited state can be approximated by a trough of cylindrical symmetry.

transitions for the lowest levels of a degenerate vibration in an E–A_1 electronic transition of a C_{3v} molecule both without and with vibronic (Jahn-Teller) splitting. While the 0–0 band is unchanged from the case of no vibronic interaction, the 1–1 band splits into three vibronic components (E–E, A_2–E, and A_1–E), similar to the three components of the 1–1 band of a Π–Σ electronic transition of a linear molecule (see Fig. 66). The 2–2 band splits into six components.

At the same time the intensity distribution in the v_k'–0 and 0–v_k'' progressions is quite anomalous. As mentioned earlier, if there is no vibronic interaction, the 0–0 band in these progressions will be by far the most prominent. But if vibronic interaction is strong, a distribution as shown in Fig. 92 arises with

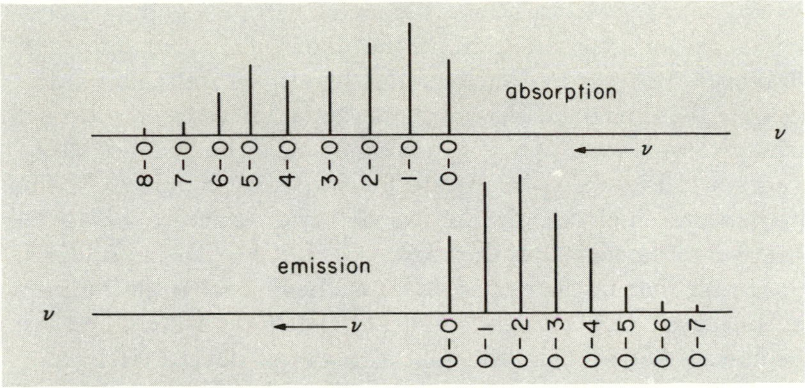

Fig. 92. Intensity distribution in a progression in a degenerate vibration for an *E–A* electronic transition of a C_{3v} molecule in emission and absorption when Jahn-Teller interaction is large [after Louguet-Higgins, Öpik, Pryce, and Sack (86), from *MM* III, p. 166).

It has been assumed that $\omega_k' = \omega_k''$. For such a case without vibronic interaction only the 0–0 band would occur.

two maxima of intensity distinctly away from the 0–0 band. In addition, the spacings in the v_k'–0 progression are irregular, an anomaly that would be even more pronounced if vibronic interaction were still larger than assumed in Fig. 92. In that case additional transitions would arise, leading to an even more complicated spectrum. No such case has yet been analyzed. It is probable that the CH_4^+ ion would have such a spectrum, and it has been suggested that the diffuse interstellar lines (which are rather irregularly spaced) represent this spectrum [Herzberg (60)].

Transitions between states with different symmetry of the equilibrium conformation. If the equilibrium conformation has different symmetry in the upper and lower state, we must apply selection rules that correspond to the common elements of symmetry.

The simplest example is a linear-nonlinear transition (that is, a transition between a state in which the molecule is linear and one in which it is nonlinear). Here, in contrast to the case of a linear-linear transition, a long progression in the bending vibration arises in absorption from the (nonlinear) ground state. Moreover, this progression includes bands with all Δv_k values, not only the even ones. However, there is a difference in subband structure for alternative v values, since according to Eq. (105) even and odd l values alternate for successive members of the progression. Such an alternation of band structure has been observed in the spectra of the radicals NH_2, CH_2, BH_2, and HCO, suggesting that in the upper states of these spectra the radicals are linear or nearly linear, while in the lower states they are nonlinear. Actually, in the first two cases, deviations from a simple vibrational formula have been found for low v', showing that the potential function of the upper state is not a simple parabola when plotted as a function of the bending coordinate x, but has a small hump at $x = 0$; that is, strictly speaking these molecules are not linear, but for vibrational levels that lie above the hump they behave as if they were linear (*quasi-linear* molecules, see p. 163). The other two radicals (BH_2, HCO) are exactly linear in the upper states of the observed spectra; there is no potential hump.

In a similar way, planar-nonplanar transitions give rise to long progressions of the out-of-plane bending vibration. The absence of such progressions in the various observed electronic transitions of CH_3 shows, according to the Franck-Condon principle, that there is very little change of bending angle between upper and lower state. Since all the upper states are Rydberg states in which the geometric structure must be the same as in the CH_3^+ ion, and since from theory there can be very little doubt that the ion is planar (there are no electrons favouring a nonplanar structure), it follows that neutral CH_3 (for which theoretical predictions cannot be made so easily) is planar—a conclusion confirmed by study of the rotational structure (see below). In further support of this conclusion, the 2100-Å system of CD_3 shows a 0–2 band in the bending vibration (ν_2), but not the 0–1 band, in agreement with the selection rules for planar-planar transitions. The value of the bending frequency derived from this assignment agrees with the ν_2 infrared band observed in a solid matrix by Milligan and Jacox (94). The conclusion that CH_3 is very nearly planar (if not planar) has also been confirmed by the investigation of the hyperfine structure of the electron-spin-resonance spectrum of CH_3 in a solid matrix [see Karplus (82)].

Forbidden transitions. No forbidden electronic transitions have been observed for any polyatomic free radicals. The most likely to be observed are those induced by vibronic interaction. These forbidden transitions have the "opposite" vibrational structure to that just described; that is, vibrational transitions with $\Delta v_k = \pm 1, \pm 3, \ldots$ occur. Further details may be found in *MM* III, p. 173f.

Isotope effect. The vibrational isotope shifts in electronic spectra of nonlinear polyatomic molecules are not as easy to predict as for diatomic molecules (see *MM* III, p. 181). Their observation, however, is often very important for an unambiguous identification of the carrier of a free-radical spectrum. The presence of an isotope shift upon isotopic substitution of a particular atom in the parent compound shows that the particular atom forms part of the radical under study. In this way the presence of one (and only one) carbon atom in the carriers of the spectra now known to be due to CH_2, NCN, HNCN, and others was firmly established. If several identical atoms are present, a 50 percent isotopic substitution leads to several isotopic bands for a given band of the normal isotope. Thus the main band of the 4050-Å group of C_3 (see p. 13), when produced in a 50:50 mixture of C^{12}:C^{13}, shows six heads instead of one head, proving unambiguously that the carrier must have three carbon atoms. Similarly, the 2160-Å band of CH_3 shows four component bands in a 50:50 H:D mixture, confirming that three hydrogen atoms are present in the molecule responsible. In a similar way the presence of two H atoms in carriers of the spectra now known to be due to CH_2 and NH_2, or two N atoms in the spectrum of NCN, was firmly established.

(4) Rotational structure of electronic transitions

Symmetric top molecules. The rotational fine structure of the electronic bands of symmetric top molecules is similar to that of the rotation-vibration bands of these molecules. The rotational *selection rules* depend on whether the transition moment of the electronic transition is parallel or perpendicular to the top axis. In the first case (|| *bands*) we have the rules

$$\Delta K = 0, \qquad \Delta J = 0, \pm 1 \tag{161}$$

with the restriction that $\Delta J = 0$ does not occur for $\Delta K = 0$. For the second case ($\perp$ *bands*) we have

$$\Delta K = \pm 1, \qquad \Delta J = 0, \pm 1. \tag{162}$$

In both cases, just as for linear molecules, certain symmetry selection rules must be obeyed—the parity rule (see p. 50)

$$+ \leftrightarrow -, \qquad + \not\leftrightarrow +, \qquad - \not\leftrightarrow - \tag{163}$$

and the rules for the overall symmetry type (see p. 51). For $\boldsymbol{C}_{3v}$ molecules these rules are

$$A_1 \leftrightarrow A_2, \qquad E \leftrightarrow E, \tag{164}$$

and for $\boldsymbol{D}_{3h}$ molecules

$$A_1' \leftrightarrow A_1'', \qquad A_2' \leftrightarrow A_2'', \qquad E' \leftrightarrow E''; \tag{165}$$

all other combinations are forbidden. For other point groups see *MM* III, p. 223. The parity rule is of importance only for nearly planar molecules when the inversion doubling is not negligibly small.

Finally, the selection rules for the $(+l)$ and $(-l)$ levels in E–A transitions must be considered (see p. 155); they are (see *MM* III, p. 224f):

$$\begin{aligned} \Delta K &= +1 \text{ occurs only for } (+l) \leftrightarrow (0), \\ \Delta K &= -1 \text{ occurs only for } (-l) \leftrightarrow (0), \end{aligned} \tag{166}$$

where (0) stands for the single levels with a certain J and K in the A state.

For transitions between nondegenerate vibronic states (A–A), if the rotational constants A and B in the upper state are nearly the same as those of the lower state, the band structure is similar to that of || *bands* in the infrared. As indicated in Fig. 93, such a band consists of a number of subbands corresponding to dif-

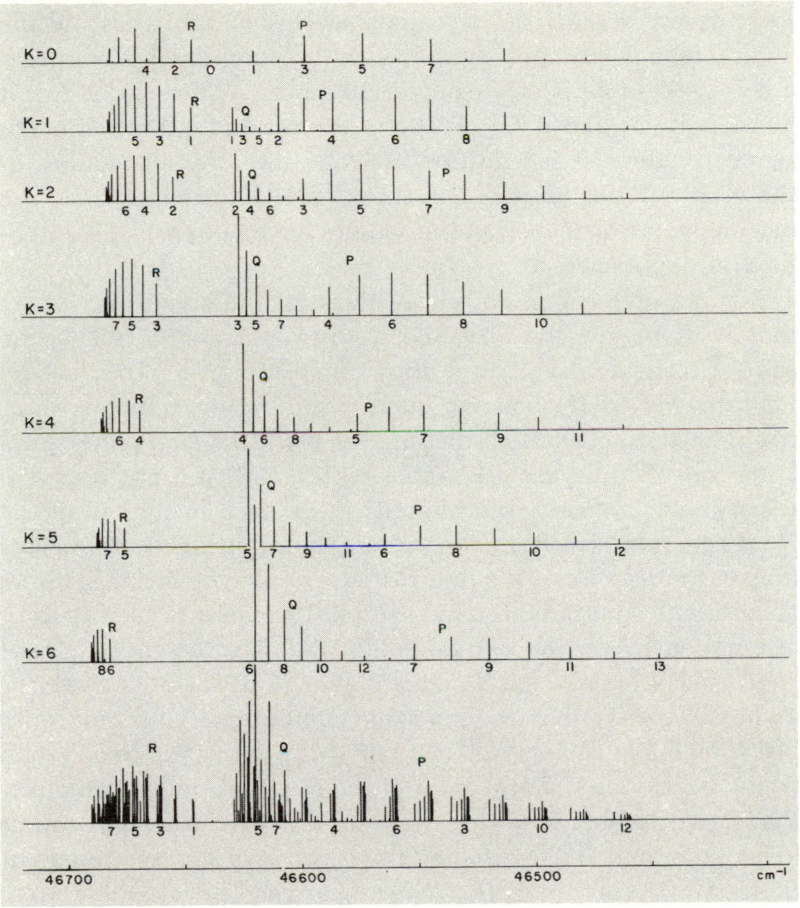

Fig. 93. Subbands of a || band of a symmetric top molecule.

At the bottom the superposition of the subbands is shown as it would appear in the spectrum. The figure is drawn approximately to scale for the 2140 Å band of CD_3 shown in Fig. 94(a). The heights of the lines indicate the intensities calculated from the Hönl-London formulae (see *MM* III, p. 226). The intensity alternation in K (see text) has been taken into account. Only the $K = 0$ subband shows an intensity alternation in J, leading to a very slight intensity alternation of the unresolved lines of the P branch in the complete band [see Fig. 94(a)].

ferent K values. Each of these has a P, Q, and R branch. For a molecule with a threefold axis of symmetry in accordance with the alternation of statistical weights mentioned earlier (p. 157), there is an *intensity alternation* of successive subbands: if the nuclear spin is zero, only subbands with K divisible by 3 appear; if it is $\frac{1}{2}$, these subbands are twice as strong as those with K not divisible by 3; and if $I = 1$, the intensity ratio is 11/8. For $\boldsymbol{D}_{3h}$ symmetry there is in addition a strong intensity alternation in the $K = 0$ subband: for $I = 0$ and $I = \frac{1}{2}$ alternate lines are missing, while for $I = 1$ alternate lines have one tenth the intensities of the other lines.

The example of Fig. 93 is drawn approximately in such a way that it represents the observed absorption band of CD_3 near 2144 Å, of which Fig. 94(a) shows a spectrogram. In this case the symmetry is $\boldsymbol{D}_{3h}$ and the nuclear spin is $I = 1$. In the spectrum, because of line broadening by predissociation, the close-lying lines of different subbands (see Fig. 93) are not resolved. A very slight intensity alternation can be seen in the tail of the P branch, and a rather strong alternation appears at the beginning of the R branch [the line $R(0)$ appears to be missing]. Such an intensity alternation can arise only because of an intensity alternation in the $K = 0$ subband, and its observation shows conclusively that the molecule is planar in at least one state, or so nearly planar that the inversion doubling is large and only one component is observed. Actually the observation of only one strong band shows, on the basis of the Franck-Condon principle, that the molecule must have the same structure in the upper and lower state and thus must be planar, or very nearly planar, in both states (point group $\boldsymbol{D}_{3h}$). Since the even lines are the weaker ones in CD_3, the transition must be of the type ${}^2A_1'$–${}^2A_2''$ or ${}^2A_1''$–${}^2A_2'$, of which only the former is compatible with the electron configuration (see Table 18).

In a molecule that is a symmetric top on account of symmetry (such as a $\boldsymbol{C}_{3v}$ or $\boldsymbol{D}_{3h}$ molecule) a $\perp$ *band* can occur only if at least one of the two electronic (or vibronic) states between which the transition occurs is degenerate. Let us for a moment disregard the first-order Coriolis splitting caused by the degeneracy.

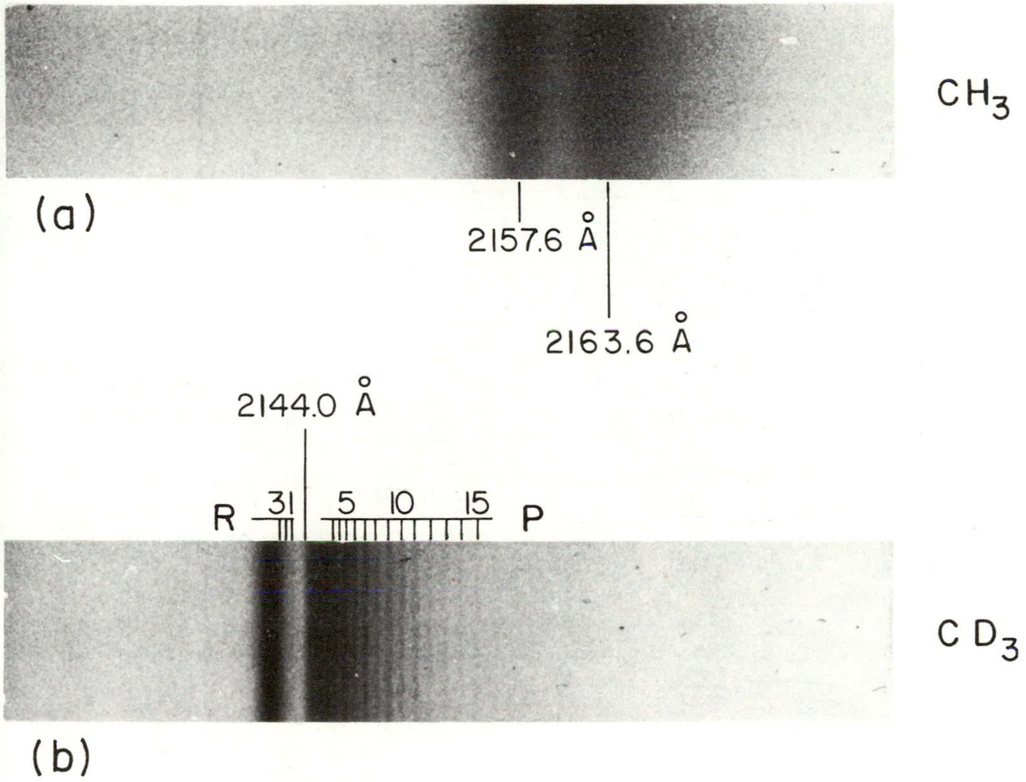

Fig. 94. Absorption bands of (a) CH_3 and (b) CD_3 near 2150 Å after Herzberg (57) (from *MM* III, p. 227).

This is a $^2A_2'$–$^2A_2''$ transition. The band structure of CD_3 is explained by Fig. 93. The resolved lines of the *P* and *R* branches are marked. Because of line broadening on account of predissociation the component lines corresponding to the different subbands are not resolved. The *Q* branch forms the head at 2144.0 Å. In CH_3, because of stronger predissociation (see p. 198), only two diffuse maxima are visible, corresponding to the *R* and $Q + P$ branches.

The energy-level diagram for this case (Fig. 95) shows immediately that, unlike the case of the || band, the subbands do not coincide even if *A* and *B* are the same in the upper and lower state. Figure 96 shows the relative positions of the subbands, as well as the resulting band structure. We see that the most prominent feature of such a band is a series of *Q* branches, which

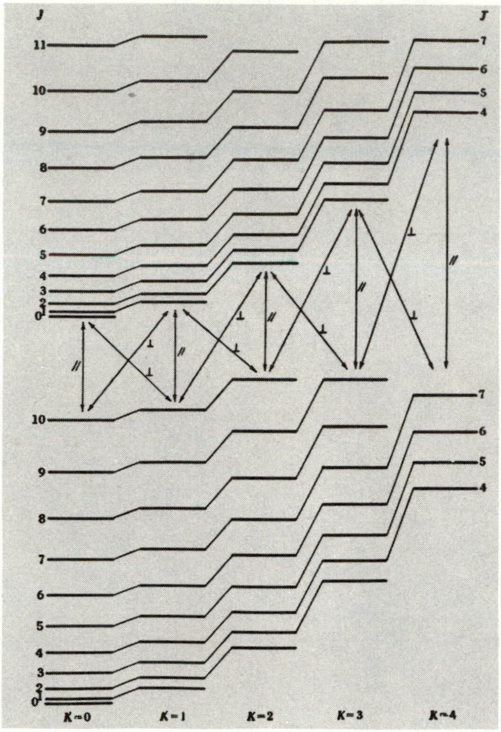

Fig. 95. Energy-level diagram showing the transitions in K corresponding to a $\parallel$ and a $\perp$ band, disregarding any Coriolis splitting that may be present (from *MM* II, p. 417).

The vertical transition arrows refer to a $\parallel$ band ($\Delta K = 0$), the oblique arrows to a $\perp$ band ($\Delta K = \pm 1$).

would be equidistant if the rotational constants A and B were the same in the upper and lower state. The spacing of the Q branches would be $2(A - B)$, as is readily seen from the energy formula (135) (neglecting the centrifugal-distortion terms). If Coriolis splitting is not neglected—that is, if $\zeta \neq 0$—we have to use the energy-level diagram given by Fig. 85 instead of Fig. 95 and take account of the selection rule (166). As is readily seen from the energy formula (142), qualitatively everything is the same as for $\zeta = 0$ (Fig. 96 still applies), but the spacing of the subbands is now

$$2[A(1 - \zeta) - B]. \tag{167}$$

The Coriolis interaction constant ζ in a general case would have both an electronic and a vibrational component (see p. 156).

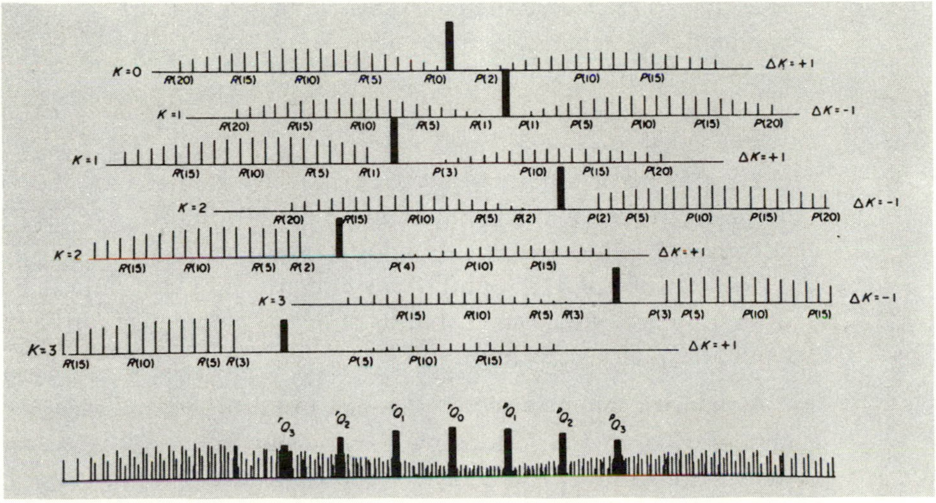

Fig. 96. Subbands of a ⊥ band of a symmetric top molecule assuming $B' \approx B''$, $A' \approx A''$ (from *MM* III, p. 230).

At the bottom the superposition of the subbands is shown. The Q branches of the subbands, in which many lines coincide, are shown as black rectangles. The heights of these rectangles and of the lines representing the P and R branches indicate the intensities. The intensity alternation that would be present in a symmetrical molecule is not indicated.

If the rotational constants A and B are not the same in the upper and lower state, the origins ν_r^0 and ν_p^0 of the subbands with $\Delta K = +1$ and -1 are given by the formula

$$\nu_{r,p}^0 = \nu_0 + [A'(1 - 2\zeta) - B'] \pm 2[A'(1 - \zeta) - B']K + [(A' - B') - (A'' - B'')]K^2, \quad (168)$$

where the $+$ or $-$ sign in front of the second square bracket refers to ν_r^0 and ν_p^0, respectively. There is now a quadratic term in K that causes the series of subbands to converge. It may be noted (see Fig. 96) that there is no zero-gap between the two branches of subbands, in contrast to the two branches of lines in a Σ–Σ band of a linear molecule.

The rotational constants B and D of the upper and lower states can be obtained from the individual subbands in the same way as the B and D values of linear or diatomic molecules (see

MM I, p. 175f.). But the constants A', A'', and ζ are not independently determinable from a $\perp$ band—at least one of the three must be known from other evidence.

In a precise evaluation of the constants we must take account of the other two centrifugal stretching constants D_{JK} and D_K (see p. 154). For prolate tops the effect of D_K is in general larger, often much larger, than that of D_J. The term in D_{JK} causes the effective B value of each subband to be slightly different.

No example of a $\perp$ band of a symmetric top free radical has been fully analyzed, but $\perp$ bands of nearly symmetric tops have been analyzed (see below).

Asymmetric top molecules. We can obtain the band structure of slightly asymmetric tops by starting out from that of a symmetric top and introducing K-type doubling (see p. 160). In place of the selection rules (161) and (162) we now have

$$\Delta K_a = 0, \pm 1, \qquad \Delta J = 0, \pm 1 \tag{169a}$$

or

$$\Delta K_c = 0, \pm 1, \qquad \Delta J = 0, \pm 1, \tag{169b}$$

depending on whether the molecule considered is close to a prolate or an oblate symmetric top. In addition to $||$ and $\perp$ bands we can now also have *hybrid bands*—bands in which both transitions with $\Delta K = 0$ and $\Delta K = \pm 1$ occur. Such bands arise when the transition moment has both a parallel and a perpendicular component, which is possible for sufficiently low symmetry—for example, for point group $\boldsymbol{C_s}$.

For *parallel bands* of slightly asymmetric top molecules we expect a doubling of all branches with $K > 0$. However, this doubling will be significant only for the lowest K values. Which component levels of the K-type doublets combine is determined by selection rules for the symmetry properties $++$, $+-$, $\cdots$ of the asymmetric top rotational eigenfunctions (see p. 162). We shall not discuss these selection rules here but refer to *MM* III, p. 245f. A good example of a $||$ band of an asymmetric top free radical has not yet been found.

For *perpendicular bands*, also, the branches in the subbands with low K values are doubled except those with $K' = 0$ or $K'' = 0$

—that is, the 0–1 or 1–0 subbands. In these latter subbands, because of the symmetry selection rules mentioned, the Q-branch transitions ($\Delta J = 0$) go to one K-type doublet component in the $K = 1$ level, the P- and R-branch transitions ($\Delta J = \pm 1$) go to the other. As a result, since the K-type splitting for $K = 1$ increases with $J(J + 1)$ and is fairly large, we may obtain a considerable difference in shading of the Q branches of the 1–0 subband from that of the 0–1 subband as well as a difference in shading between Q and P, R branches in each of these subbands.

As an example, Fig. 97 shows a spectrogram of the 3400-Å band of the HNCN radical. The wide spacing of the Q branches shows that A is large; that is, the moment of inertia about the top axis is very small. This is possible only if the three heavy atoms lie almost on a straight line, with the H atom off the line. The two large moments of inertia are nearly equal, and so we obtain the typical structure of a perpendicular band of a symmetric top (compare Fig. 96). From the energy formula (147) with $W_{J_\tau}{}^v \approx K^2$ we see that the spacing of the subbands is approximately $2[A - \frac{1}{2}(B + C)]$. Closer inspection of the spectrogram, Fig. 97, reveals that the Q branches of the 0–1 and 1–0

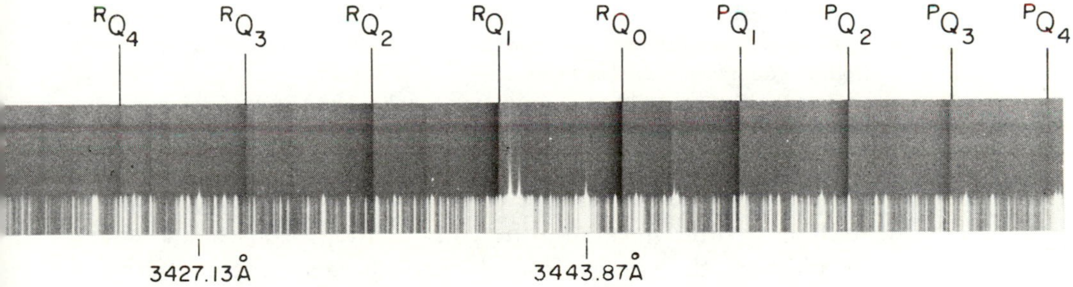

Fig. 97. The 0–0 band of the $\tilde{A}$–$\tilde{X}$ system of the HNCN radical at 3440 Å in absorption after Herzberg and Warsop (72) (from *MM* III, p. 258). [Reproduced by permission of the National Research Council of Canada from the *Canadian Journal of Physics.*]

The heads of the Q branches of the subbands are marked; the heads of the P branches are much weaker but recognizable to the right of the Q heads of several subbands. Note that because of the K-type splitting no P head is visible for the 0–1 subband (marked $^PQ_1$).

subbands show a great difference in shading. Moreover, measurements at higher resolution reveal the expected doubling of the branches in the 1–2 and 2–1 subbands [Herzberg and Warsop (72)].

As a second example, Fig. 98 shows one of the subbands (3–2) of the HNO radical. Here, because of the greater asymmetry, the *K*-type doubling of each of the branches is clearly exhibited. Similar structures have been observed for HCF [Merer and Travis (92)], HSiCl, and HSiBr [Herzberg and Verma (71)]. The number of missing lines at the beginning of the branches of each subband immediately establishes the *K* values.

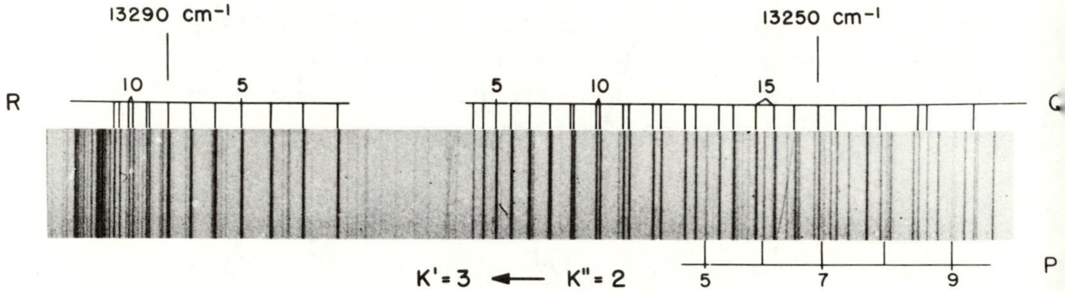

Fig. 98. The 3–2 subband of the 000–000 band of the $\tilde{A}$–$\tilde{X}$ system of the HNO radical at 7534 Å in absorption, after Dalby (26) (from *MM* III, p. 259). [Reproduced by permission of the National Research Council of Canada from the *Canadian Journal of Physics.*]
Note the *K*-type doubling (asymmetry doubling) at higher *J* values.

In strongly asymmetric tops, *K* is no longer a good quantum number and therefore the selection rules for *K* no longer apply; subbands with ΔK values other than 0 and ± 1 arise, and since, in addition, the energy formula is no longer simple, the band structure is much more complicated than for nearly symmetric tops. An intermediate case is provided by the $\tilde{A}^1B_1$–$\tilde{X}^1A_1$ transition of the CF_2 radical, whose rotational structure was analyzed by Mathews (89).[4] No free-radical spectrum for which in both

[4] Note that in *MM* III, p. 510 and 603 this transition was provisionally considered to be $^1A_1$–$^1A_1$. More detailed analysis by Mathews (89) has since led unambiguously to the assignment $^1B_1$–$^1A_1$.

upper and lower state the system is a strongly asymmetric top has yet been analyzed. We shall therefore not discuss further details (see *MM* III, p. 261f).

Linear-bent transitions. Several cases of electronic transitions are known between an upper state in which the molecule is bent and a lower state in which it is linear (that is, bent-linear transitions), but none of these refers to free radicals. On the other hand, many examples of linear-bent transitions[5] have been observed for free radicals. In the bent conformation the molecule is in general an asymmetric top; therefore the asymmetric-top selection rules apply. If the asymmetry is small, the symmetric-top selection rules

$$\Delta K = 0, \pm 1, \qquad \Delta J = 0, \pm 1 \tag{170}$$

may still be used. The quantum number K in the upper "linear" state is l (see p. 92).

As already mentioned, in a linear-bent transition, according to the Franck-Condon principle, the bending vibration (ν_2' for a triatomic molecule) is excited in the upper state. For each v_2' several subbands arise corresponding to different l_2 values. In the progression of bands corresponding to the bending vibrations there is an alternation of even and odd l_2. Except for this restriction of the l_2 values, the subband structure is entirely similar to that of parallel or perpendicular bands of symmetric top molecules, since the energy in the upper state depends on l_2 in the same way ($g_{22}l_2^2$) as it depends in the lower state on K [namely, $(A'' - \frac{1}{2}(B'' + C''))K^2$; see p. 160].

The subbands have a quasi-diatomic structure; that is, for $||$ bands they are of the type Σ–Σ, Π–Π, or Δ–Δ, . . . , depending on l_2, while for $\perp$ bands they are of the type Σ–Π, Π–Δ, . . . , Π–Σ, Δ–Π, ... Unlike the behavior of the subbands of symmetric top molecules, there is here in general a fairly large K-type (asymmetry) doubling in the lower state, which shows

[5] In these short-hand designations we follow the established usage mentioned earlier (p. 49) that the upper state is put first. Thus in linear-bent transitions the molecule is linear in the upper and bent in the lower state.

itself either directly as a line splitting or indirectly, for Σ–Π or Π–Σ subbands, as a combination defect.

As an example, let us consider the absorption spectrum of the HCO free radical. Here a progression of quasi-diatomic bands with a spacing of about 1500 cm^{-1} was observed. Figure 99, a

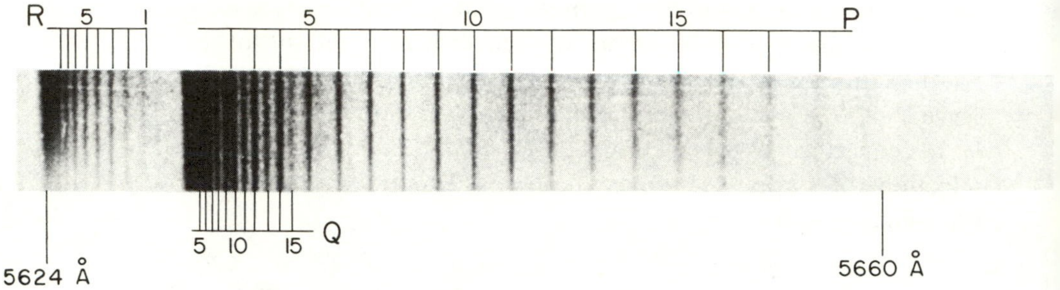

Fig. 99. The 0 11 0–000 band of the $\tilde{A}$–$\tilde{X}$ system of HCO at 5624 Å. Only the 0–1 subband is observed. The three branches *P*, *Q*, and *R* are marked. They are numbered according to the *N* values in the lower state. Each line is an unresolved doublet.

spectrogram of one of the bands obtained, clearly shows a *P*, *Q*, and *R* branch. The doublet splitting corresponding to $S = \frac{1}{2}$ is not resolved, because the lines are intrinsically broad. On the basis of the simple band structure one might be inclined to conclude that the radical is linear in both upper and lower state. However, a large combination defect between *P*, *R*, and *Q* branches was found,[6] which can only be explained as an asymmetry doubling in the lower state; that is, we must conclude that the molecule is bent in the lower state and thus is a nearly symmetric top. However, once this conclusion was accepted, another puzzle remained: why did only one subband of each vibrational transition appear? The only way of accounting for this observation seemed to be the *ad hoc* assumption that in the upper state all levels with $l_2 > 0$ are absent on account of predissociation. If this were true, the observed bands would be bands of alternate v_2 values. The bands with intermediate v_2

[6] Similar to Λ-type doubling (p. 82); for more details see *MM* III, p. 194f.

values would have $l_2 = 1, 3, \ldots$ and would be too diffuse to be observed at high resolution. Indeed, when spectra were taken at low resolution, bands with these intermediate v_2 values were found as diffuse features approximately halfway between the sharp bands. At the same time, it was found that the sharp bands are accompanied by very broad features corresponding to $l_2 = 2$. Thus, the *ad hoc* assumption was completely confirmed.

The K-type doubling in the lower state of the HCO bands gives the value of B–C, while the effective B value gives $\frac{1}{2}(B + C)$ (see p. 160). The resulting values of B and C are not sufficient to determine the structure of the radical in its ground state; however, the observation of the corresponding spectrum of DCO yields two further rotational constants from which, together with the constants for HCO, the geometrical data of the radical in the ground and excited state can be evaluated. Fig. 100 shows the resulting structures with Ogilvie's (105) corrections for the lower state.

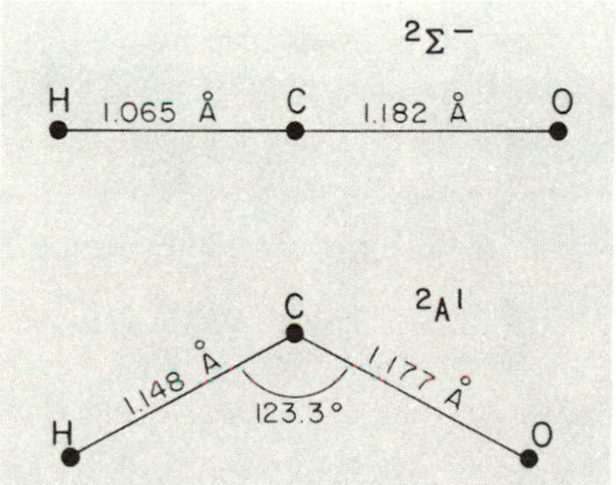

Fig. 100. Observed geometrical structure of the HCO radical in the $\tilde{A}\ ^2\Sigma^-$ and $\tilde{X}\ ^2A'$ states.

According to Table 10 the ground state of linear HCO is predicted to be $^2\Pi$. Such a state is subject to Renner-Teller interaction. If we assume that this interaction is strong and leads to one state with a bent and one with a linear equilibrium conformation in the way shown in Fig. 56(b) we can account for the observed transition as the one between the two component states. On this basis we would expect the transition moment of the electronic transition to be perpendicular to the plane of the molecule. This expectation is indeed verified by the observation that the Q lines correspond to the lower K-doublet components of the ground state (see *MM* III, p. 198). If we assume that the ground state is $^2A'$, it follows that the upper state is $^2A''$—or rather, since the molecule is linear in this state, $^2\Sigma^-$.

The absorption spectrum of BH_2 presents another good example of a linear-bent transition. Here several subbands are observed for each value of v_2', and these subbands show a clear alternation between even and odd l_2 values in the upper state. Figure 101 shows two subbands. For several lines the very small spin doublet splitting is clearly resolved. The spectrogram also shows a weaker isotope band corresponding to $B^{10}H_2$, which, together with the intensity alternation in the branches, confirms the assignment to BH_2. Because of the symmetry of BH_2, the geometrical data can be obtained from the spectrum of BH_2 alone; they are confirmed by those of BD_2. These data are included in Fig. 75.

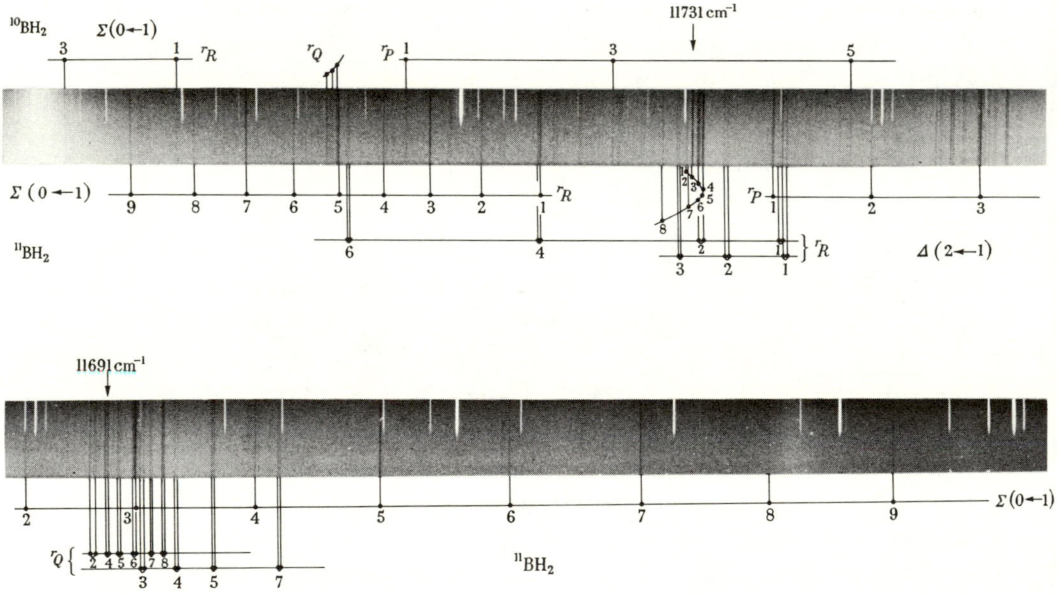

Fig. 101. The 0–1 and 2–1 subbands of the BH_2 absorption band near 8520 Å after Herzberg and Johns (64).

The assignments of the lines of $^{10}BH_2$ and $^{11}BH_2$ have been indicated above and below the spectrogram.

The discovery of the spectra of NH_2 and CH_2 in the red region preceded that of BH_2. For NH_2 and CH_2 (singlet) the subband structure is not nearly so obvious as in BH_2 because in the lower state these radicals are strongly asymmetric tops. That is why the analysis of the NH_2 spectrum took many years [Dressler and Ramsay (38)]. For both NH_2 and CH_2 an alternation of subband structure in successive bands of the v_2' progression was recognized even before a full analysis was accomplished. This alternation shows clearly that in the upper state both radicals are effectively linear. However, in both cases deviations from a simple vibrational formula have been found for the lowest v_2' values, indicating the presence of a small potential hump corresponding to the linear conformation. All of the observed vibrational levels are above this hump. The equilibrium configurations for NH_2 and singlet CH_2 are shown in Fig. 75. For NH_2, if it were linear, according to Table 9, the ground state would be $^2\Pi$. Strong Renner-Teller interaction causes a splitting of this $^2\Pi$ state into two states, $^2A_1$ and $^2B_1$, of which the latter has a bent equilibrium conformation (see the similar discussion for HCO above). For linear CH_2 the ground state is $^3\Sigma_g^-$ and the lowest singlet state is $^1\Delta_g$. We must assume that the two observed singlet states of CH_2 ($^1A_1$ and $^1B_1$) are derived from $^1\Delta_g$ of the linear conformation.

V. DISSOCIATION, PREDISSOCIATION, AND RECOMBINATION

A. CONTINUOUS SPECTRA

As mentioned earlier, in diatomic molecules a continuous range of energy levels (corresponding to dissociation or recombination) joins onto every series of vibrational levels (and even arises for those electronic states that have no discrete vibrational levels). Correspondingly continuous absorption spectra are observed joining onto the progressions of discrete bands corresponding to various excited electronic states. The continua will be observable only if there is a fairly large change of the equilibrium internuclear distance r_e in going from the lower to the upper state, since then the Franck-Condon maximum will be at high v', or even in the continuum (see Fig. 36 and the accompanying discussion).

There are only two cases of diatomic free radicals, SO and ClO, where the convergence of a v' progression has been observed, and the adjacent continuum is seen (see p. 69 and Fig. 39). The convergence limit supplies a precise value for a dissociation limit of the radical (see below).

If the shift in r_e is large, or if the upper state potential function has no minimum, a continuum without preceding bands may be observed (as, for example, in F_2), but such cases have not yet been identified for diatomic free radicals.

The number of vibrational degrees of freedom (normal coordinates) in a linear polyatomic molecule is $3N - 5$, in a nonlinear polyatomic molecule it is $3N - 6$. Even for a triatomic molecule ($N = 3$) this number is larger than 2; thus we must consider multidimensional potential surfaces. Only if the vibrational motion in upper and lower state is one-dimensional, as in a diatomic molecule, will a simple progression of bands be observed in absorption. While many progressions of this type have been observed for various polyatomic molecules and radicals, in no case have convergence limits of such progressions been found nor the continua joining onto them. However, many examples of continuous absorption spectra without preceding convergence limits are known for stable polyatomic molecules, but very few for free radicals, mainly because of the difficulty of identification. Here we may note that in each electronic state of a polyatomic molecule there are several dissociation limits corresponding to different dissociation products; for example, NCO can dissociate into N + CO or NC + O or N + C + O.

Besides continua corresponding to *dissociation*, there are also continua corresponding to *ionization*. Only very few Rydberg series (see *MM* I, p. 322) of free radicals have been observed to fairly high members (for example, those of CH, CH_2, and CH_3), but in no case was the absorption strong enough that the adjoining continuum could be seen.

The limit of the Rydberg series gives an ionization potential of the radical, but not necessarily the lowest ionization potential. For CH_2, for example, a Rydberg series is observed in absorption from the $^3\Sigma_g^-$ ground state in which CH_2 is linear (see Fig. 75). Since only 0–0 bands are observed, it appears certain that the limit of the Rydberg series corresponds to a linear conformation of CH_2^+. On the other hand, in analogy to BH_2 it is very likely that CH_2^+ in its ground state is nonlinear, having an energy less than that of the linear conformation (see Fig. 75 for BH_2).

Continuous *emission* spectra of molecules and radicals are even more difficult to assign unambiguously than continuous absorption spectra. Such spectra arise by transitions from a stable upper state to an unstable lower state, like the well-known H_2 continuum, for which the lower state corresponds to dissociation into two normal H atoms.

Emission continua corresponding to recombination will be discussed briefly in section C.

B. DIFFUSE SPECTRA: PREDISSOCIATION AND PRE-IONIZATION

(1) Auger processes

If for any atomic system the approximate solution of the wave equation leads to two accidentally close-lying energy levels E_1

and E_2 of the same total angular momentum and the same symmetry [see Fig. 102(a)], the introduction of higher approxi-

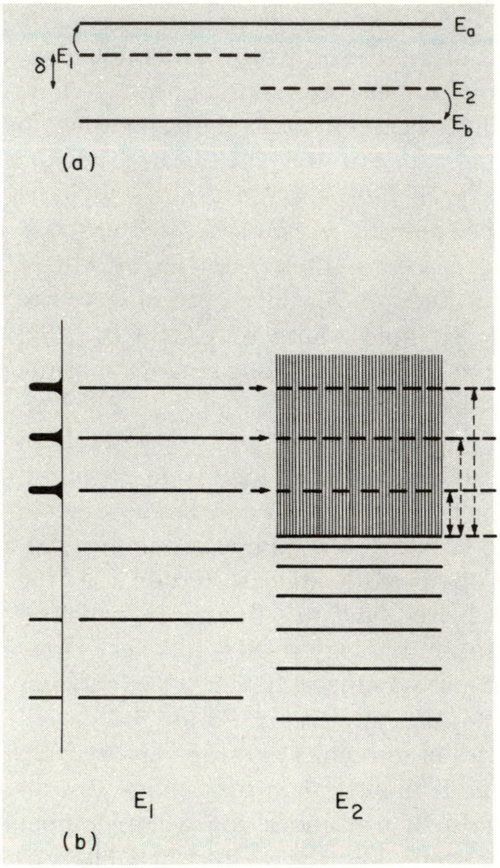

Fig. 102. Energy-level diagrams explaining (a) perturbations (b) Auger processes.

In (a) the broken lines represent the zero-approximation levels, the full lines the actual levels after higher-order terms have been taken into account. In (b) the three uppermost levels of the E_1 series are overlapped by the continuum of the E_2 series, leading to broadening of these levels as indicated at the extreme left. The radiationless transitions are indicated by the horizontal arrows. The vertical double arrows at the right indicate the kinetic energy of the dissociation (or ionization) products.

mations will lead to a shift of the energy levels away from each other. The corrected levels E_a and E_b are given by

$$E_{a,b} = \tfrac{1}{2}(E_1 + E_2) \pm \tfrac{1}{2}\sqrt{4(W_{12})^2 + \delta^2}, \qquad (171)$$

where δ is the "original" energy difference $E_1 - E_2$ and W_{12} is a measure of the interaction between E_1 and E_2 brought about by the higher terms in the Hamiltonian which were originally neglected. The eigenfunctions of the two states E_a and E_b are mixtures of the zero-approximation eigenfunctions ψ_1 and ψ_2, corresponding to E_1 and E_2, respectively:

$$\psi_a = c\psi_1 - d\psi_2, \qquad \psi_b = d\psi_1 + c\psi_2, \tag{172}$$

where

$$\left.\begin{matrix} c \\ d \end{matrix}\right\} = \sqrt{\frac{\sqrt{4|W_{12}|^2 + \delta^2} \pm \delta}{2\sqrt{4|W_{12}|^2 + \delta^2}}}. \tag{173}$$

If $\delta \approx 0$ or $W_{12} \gg \delta$, $c \approx d \approx 1/\sqrt{2}$; that is, we have a 50:50 mixture. If, as is usual, the two levels E_1 and E_2 belong to two different series of levels, the interaction will cause a deviation in each series from a smooth formula—that is, a *perturbation*. Such perturbations are fairly often found in the rotational and vibrational series of diatomic and polyatomic molecules. In the vibrational series of the latter they arise even in the ground electronic state, when no other electronic state is nearby, simply between vibrational levels belonging to different vibrations; they are known as *Fermi resonances* (see p. 92).

If one of the two levels, say E_2, belongs to a continuous range of levels corresponding to dissociation or ionization, all levels of the set E_2 lying in the neighbourhood of E_1 may perturb E_1—some shifting it down, some shifting it up; and as a result we obtain a slightly *diffuse* level in place of E_1, as indicated in Fig. 102(b). The mixing of the eigenfunctions of the two states implies that if the system is brought to E_1 it will soon adopt properties of E_2; that is, it will dissociate or ionize. In an approximate way we can express the situation by saying that a *radiationless transition* takes place from the discrete state into the continuum (at equal energy) leading to decomposition. Such processes are called *Auger processes*, after Auger who first discovered them in the X-ray region: he found that one X-ray photon may lead to the emission of two photoelectrons, one emitted directly by an ordinary photoelectric effect (for example, from the K shell) and another emitted immediately afterward by way of such a radiationless transition (since the K level reached in the first stage is overlapped by the continuum corresponding to the removal of an L electron from the ion).

Auger processes have been found to be of great importance

not only in the X-ray region but also in the optical region of atomic spectra, of molecular spectra, of solid-state spectra, in nuclear physics, and even in elementary particle physics. It must be emphasized that the concepts of Fermi resonance (or perturbations) and Auger process are constructs that would be superfluous if one could work at all times with exact solutions of the rigorous wave equations of the systems considered here. Yet for an understanding of the observed phenomena and their simplified description these concepts are of great importance.

The nature of the radiationless transition process described above leads us immediately to the following experimentally verifiable characteristics, which may be used as criteria for its existence in a given case:

(1) *Decomposition* of the system following excitation to the excited state that is subject to an Auger process. This is the characteristic by which the process was discovered: the emission of a second photo-electron (Auger-electron) after production of the highly excited K state of the ion by the absorption of an X-ray photon.

(2) *Broadening* of those absorption lines whose upper states are subject to an Auger process. Such broadening was discovered only much later in the X-ray region but was the basis of the discovery of Auger processes in the optical region for both atoms and molecules (see below).

(3) *Weakening of the emission lines* from levels subject to Auger effect compared to emission from similar levels of similar energy not subject to Auger effect. Such weakening will occur when the radiationless transition probability into the continuum is comparable to or larger than the radiative transition probability to lower states.

(4) Occurrence of *resonances* in the collisions of appropriate particles (atoms, molecules, electrons, and so on) leading by *inverse Auger processes*—that is, by radiationless transitions—from the continuum to the discrete levels. Such processes have been studied a great deal in nuclear and elementary particle physics, and more recently in the scattering of electrons by atoms and molecules.

If a direct transition from the ground state to the continuum is possible as well as to the "discrete" levels overlapped by it, an interference phenomenon between the two transitions occurs which causes the corresponding absorption lines to be asymmetric: on one side of the original line there is reduced absorption (an *apparent emission* wing); on the other there is increased absorption making the *line shape asymmetric*. Fano (41) has calculated these shapes for different ratios q of discrete and continuous absorption; his curves are reproduced in Fig. 103. Such curves also

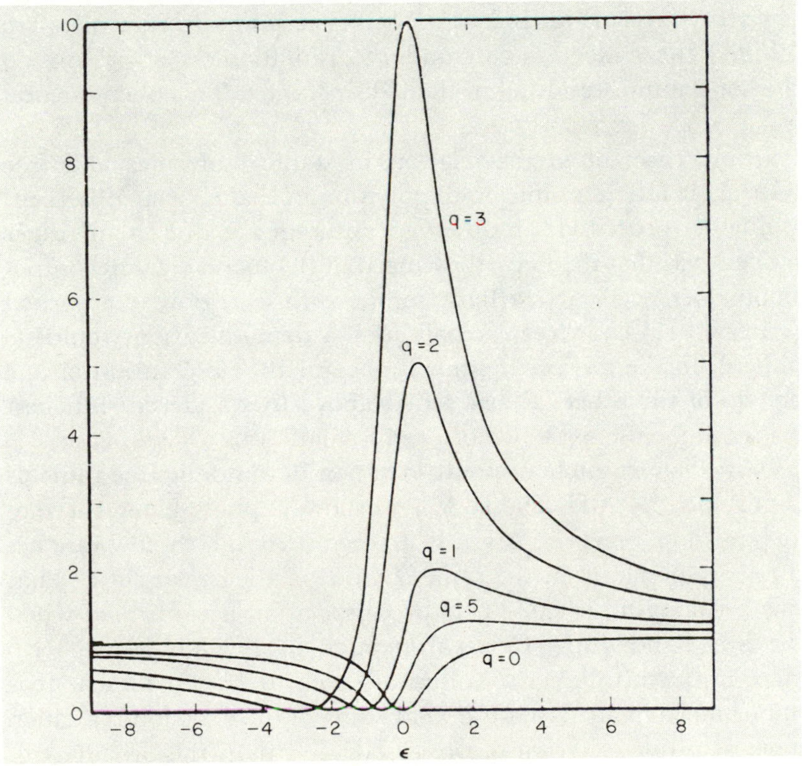

Fig. 103. Theoretical line shapes of absorption lines whose upper state is subject to Auger effect, after Fano (41).

The line shape is expressed in terms of the continuous absorption coefficient, which is put equal to 1. The shape parameter q is the ratio of the unperturbed discrete and continuous transition moments. The abscissa scale is an energy scale in units of the halfwidth.

apply to the scattering of electrons or other particles near a resonance in an inverse Auger process.

(2) Predissociation

The occurrence in molecules of Auger processes corresponding to dissociation was first recognized by Bonhoeffer and Farkas (10) after the observation of genuine diffuseness in certain molecular spectra by V. Henri (49). In molecules it often happens that discrete excited states lie above the first dissociation limit and therefore are overlapped by a continuous range of energy levels [Fig. 102(b)]. Thus if certain selection rules are fulfilled the molecules can undergo radiationless transitions to the continuum, resulting in their dissociation. This phenomenon is called *predissociation*.

Many cases of predissociation in stable molecules and free radicals, both diatomic and polyatomic, have been observed. While the proof that an observed diffuseness is due to an Auger process was first made by showing that photochemical decomposition does occur when light in the diffuse region is absorbed [criterion (1)], for free radicals such a demonstration would be difficult in the gas phase and in general the observation of one or two of the other criteria listed above is considered sufficient evidence for the existence of predissociation in a given case.

A striking example of predissociation in diatomic free radicals is provided by AlH. Figure 104(a) shows a photometer curve of an emission band of AlH. It is seen that all three branches suddenly break off at the same J value of the upper state. That this breaking-off is caused by predissociation is confirmed when the same band is observed in absorption, as shown in Fig. 104(b). Here lines with higher J values are seen to be broadened. It is important to realize that the weakening of the emission is a much more sensitive criterion of predissociation than the broadening, since for a noticeable broadening to occur the line width must be larger than about 0.1 cm^{-1}, which is 100 times the natural line width, implying that the radiationless transition probability γ must be 100 times the radiative transition probability β, while a 50 percent weakening will arise if $\gamma = \beta$. This difference is the

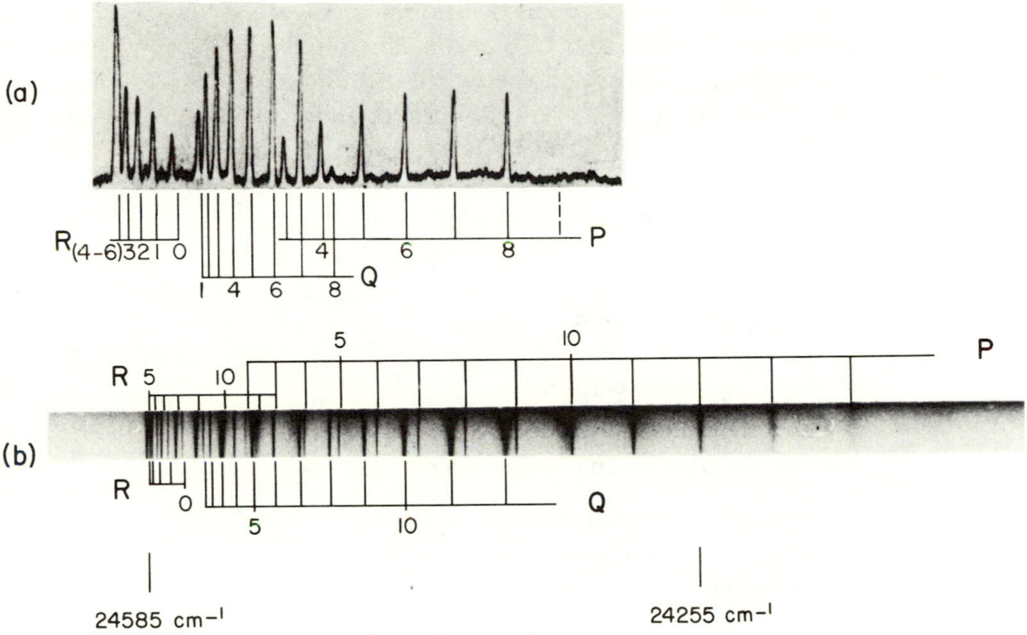

Fig. 104. Predissociation in the 1–1 band of the AlH radical (a) breaking-off in emission [after Bengtsson-Knave and Rydberg (8)] (b) diffuseness in absorption.

The J values given refer as usual to the lower state; if $\Delta J = +1, 0, -1$ is added for R, Q, and P branches, respectively, we see that breaking off occurs for the same J' (= 7) in all three branches. Broadening in absorption becomes noticeable only at somewhat higher J' values.

reason that the absorption band of AlH [Fig. 104(b)] shows broadening only at somewhat higher J values than those at which breaking-off in emission is observed. Another example of predissociation is provided by the CH band shown in Fig. 49, p. 85.

For AlH inverse predissociation has been observed as well. If Al and H atoms are brought together, a chemiluminescence arises in which only the lines with $J' > 7$ are emitted both in the laboratory [Stenvinkel (123)] and in the spectrum of the star χ Cygni [Herbig (50)]. Clearly here the levels with $J' > 7$ are formed by inverse predissociation followed by emission, and thus stabilization of the molecule.

A good example of predissociation in a polyatomic free radical is provided by CH_3 and CD_3 and is illustrated by Fig. 94. There is a striking difference in the width of the lines for the two isotopic radicals: for CH_3 no individual rotational lines are recognizable; for CD_3 they can be recognized although they are still fairly broad.

The only case of a polyatomic radical (or molecule) for which both a breaking off in emission and a broadening in absorption has been observed is HNO [Clement and Ramsay (19), Bancroft, Hollas, and Ramsay (5)].

(3) Types of predissociation

Corresponding to the three forms of energy of a molecule one may distinguish three cases of predissociation.

Case I: predissociation by electronic transition,
Case II: predissociation by vibration,
Case III: predissociation by rotation.

In case I predissociation the radiationless transition takes place between the discrete levels of one electronic state and the continuum of another; in cases II and III there is no change of electronic state. Rather, in case II a transition to the continuum associated with a different vibration occurs, and in case III from the higher rotational levels of a stable vibrational level to the continuum of the same vibration. Case II cannot arise for diatomic molecules but is of considerable importance for polyatomic molecules, radicals, and ions. Most unimolecular decompositions belong to this case. Case III has been observed only for diatomic molecules and is not likely to be of importance for polyatomic radicals. Both cases II and III can be considered in a semiclassical way by following the motion of the molecule in a (multidimensional) potential diagram.

(4) Selection rules for predissociation

In a first approximation, for an Auger process to occur, the matrix element of the perturbation function between the discrete and the continuous state

$$\int \psi_1^* W \psi_E \, d\tau \tag{174}$$

must be different from zero. Here W consists of the terms in the Hamiltonian neglected in deriving the zero-order energy levels, and ψ_1 and ψ_E are the zero-order eigenfunctions of the discrete and continuous range. On this basis it is readily seen, since the Hamiltonian is totally symmetric, that the overall symmetry properties of the two states must be the same and that the total angular momentum must be the same. Here it should be noted that the symmetry properties as well as J are well defined also in a continuous range of energy levels.

Thus we have for all types of molecules

$$\Delta J = 0 \tag{175}$$

and

$$+ \not\leftrightarrow -, \qquad + \leftrightarrow +, \qquad - \leftrightarrow -. \tag{176}$$

In addition, we have for symmetrical diatomic or linear polyatomic molecules

$$s \not\leftrightarrow a, \qquad s \leftrightarrow s, \qquad a \leftrightarrow a \tag{177}$$

and similarly

$$A_1 \leftrightarrow A_1, \quad \ldots, \quad E \leftrightarrow E, \quad \ldots \tag{178}$$

for the overall species of nonlinear molecules.

If spin-orbit interaction is weak, we have in addition

$$\Delta S = 0; \tag{179}$$

that is, singlet-triplet or doublet-quartet radiationless transitions are very weak compared to singlet-singlet, doublet-doublet, . . . transitions.

If the overall eigenfunction can be represented as a product $\psi_e \psi_v \psi_r$ [see Eq. (41)], the matrix element (174) can be resolved into three factors, one corresponding to the electronic, one to the vibrational, and one to the rotational eigenfunctions. As a consequence, in this approximation, predissociations are allowed only between electronic states of the same species (*homogeneous predissociations*). Similarly, the vibrational eigenfunctions of the two states must have the same species, and so must the rotational eigenfunctions.

Even if the symmetry requirement for the vibrational component of the matrix element (174) is fulfilled, its value will still be small unless the vibrational eigenfunctions of the two states overlap sufficiently; in other words, we have to apply the Franck-Condon principle (see p. 69f.) if we want to predict the vibrational levels for which the radiationless transition probability is a maximum. The result is that predissociation will be strong only if the potential surfaces (or curves for diatomic molecules) of the two states intersect or come close together.

Predissociation does not necessarily occur (or become observable) as soon as the energy of the discrete levels is higher than the lowest dissociation limit. An absence of predissociation above a limit is frequently a consequence of the Franck-Condon principle. Consider, for example, the case in which the two potential functions of the two electronic states involved have the form sketched in Fig. 105. At the level *BC* of the lower dissociation limit the two curves (or surfaces) are widely separated, and therefore, according to the Franck-Condon principle, predissociation has a very low probability, while near the point of intersection *E*, which lies in this example substantially above the limit, predissociation has a high probability. Slightly below the point of intersection, since the vibrational wave functions still overlap substantially, predissociation can still occur; another way to express this state of affairs is to say that predissociation occurs on account of "tunneling."

Tunneling depends strongly on the reduced mass of the vibration, and that is why we often observe that diffuseness is much less pronounced in deuterated molecules and radicals than in the corresponding normal molecules. Striking examples of this effect can be seen in the spectra of CH_2, CD_2 and CH_3, CD_3 in Fig. 9 and Fig. 94. Indeed, the structure of these radicals could not have been established if it had not been for the sharpening of the lines for the deuterated species. It was found that even the small difference in reduced mass between $C^{12}H_2$ and $C^{13}H_2$ produces a noticeable difference in diffuseness.

As an example of the application of the selection rules let us consider the predissociation of CH_3 in the 2160-Å band (Fig. 94), whose upper state is a $^2A_1'$ state assuming $\boldsymbol{D}_{3h}$ symmetry. The

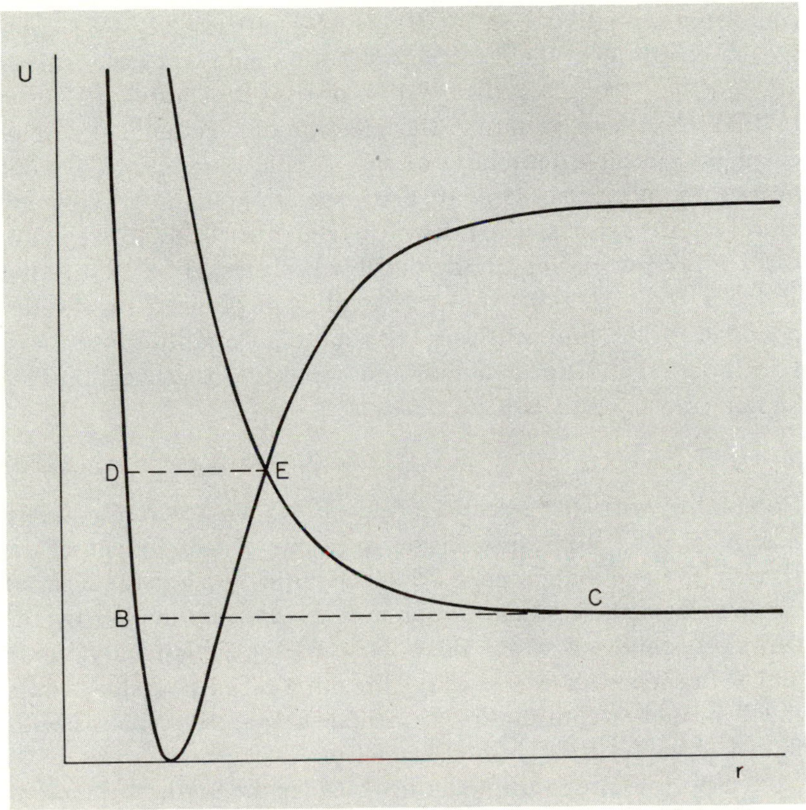

Fig. 105. Potential functions of two electronic states between which a radiationless transition (predissociation) can occur.

The diagram applied to polyatomic molecules gives the potential energy as a function of one coordinate only.

state causing the predissociation is a state of the complex $H + CH_2$ which presumably has $\boldsymbol{C}_{2v}$ symmetry. Since according to the previous selection rules the electronic species of the discrete and the continuous state must be the same, the complex must arise in a $^2A_1$ state (the equivalent of $^2A_1'$ in $\boldsymbol{D}_{3h}$ symmetry) —that is, the CH_2 radical produced by the predissociation must be in the $^1A_1$ lower state of the red CH_2 bands (see page 189). A $^3A_1$ state would also be compatible with the selection rules, but there is no low-lying $^3A_1$ state in CH_2. On the other hand, the ground state $^3\Sigma_g^-$ of CH_2 correlates with $^3B_1$ in the nonlinear conformation—that is, with $^2B_1$ (or $^4B_1$) of the $CH_2 + H$ complex. According to the electronic selection rule this state cannot arise in the (strong) predissociation of the $^2A_1'$ state of CH_3.

Predissociations violating the electronic selection rules may occur if vibronic interactions or ro-vibronic interactions are not negligible. While no cases of predissociation made possible by vibronic interactions have been clearly identified, there are many cases of predissociations made possible by interaction of rotation and electronic motion; they are readily recognized by the dependence of the line width on the rotational quantum numbers. For diatomic and linear polyatomic molecules the selection rule for such *heterogeneous predissociations* is

$$\Delta\Lambda = \pm 1. \tag{180}$$

Combining with the selection rule (176), we see immediately that in a $\Pi \rightarrow \Sigma$ predissociation only one Λ-component of the Π state (the one that for each J has the same parity as the Σ state) can be affected. An example of such a case is provided by the $C^2\Pi$–$X^2\Sigma$ bands of MgH near 2430 Å, for which only the P and R branches show a striking breaking-off in emission, while the Q branches continue in a normal way; indeed, bands with higher v' values have only Q branches.

As a polyatomic example let us consider the predissociation of the HCO free radical. As already mentioned (p. 187), in the upper $^2A''$ state of the red absorption system all bands with $K' \neq 0$ are diffuse, while the bands with $K' = 0$ are sharp. It will be recalled that in the upper state the molecule is linear and that both upper and lower state arise from a $^2\Pi$ state of the linear conformation on account of strong Renner-Teller interaction. For $K' = 0$ the upper states are vibronically Σ^-, while the continuum belongs to the $^2\Sigma^+$ (or $^2A'$) state arising from normal atoms and cannot cause the predissociation of $^2\Sigma^-$. But for $K' > 0$ there are levels with $K = K'$ and the correct symmetry in the continuum; thus predissociation is allowed and indeed observed to be strong. In the discrete bands (with $K' = 0$) a slight diffuseness increasing with J has been found; it is probably caused by a heterogeneous predissociation of the $K' = 0$ ($^2\Sigma^-$) levels by the $K = 1$ levels of the continuum.

(5) Metastable ions

The predissociation of molecular ions in long-lived excited states is responsible for the occurrence of broad peaks at non-integral mass values in a mass spectrometer. In mass spectrometry these ions are usually referred to as metastable ions, but the essential point for their observation is that after their normal acceleration they predissociate before entering the analyzing magnetic field; that is, they are *predissociating metastable ions*. In order to be observable they must have a long lifetime both for the radiative transition to a lower state and the radiationless transition leading to decomposition (predissociation).

An interesting example is observed for H_2S. If this molecule is bombarded by electrons, normal H_2S^+ ions are formed at an accelerating voltage above the ionization potential of 10.47 eV. At energies above 13.1 eV S^+ ions are observed, some of which do not occur at their regular position ($m = 32$) but at an apparent mass $m = 30.1$, from which it can be concluded that they were accelerated as H_2S^+ ions, which subsequently spontaneously decomposed (predissociated) into $S^+ + H_2$ [Dibeler and Rosenstock (30)]. The state that causes the predissociation is a quartet state (since the ground state of S^+ is ^4S and that of H_2 is $^1\Sigma_g^+$), while the state formed by ionization of H_2S can only be a doublet state. The need for breaking the spin rule ($\Delta S = 0$) makes the predissociation sufficiently slow that dissociation occurs only after acceleration of the H_2S^+ ion. At the same time the reason for the metastability of the excited H_2S^+ ion (before predissociation) must be the fact that it is only vibrationally excited, and infrared emission (while not forbidden) is sufficiently slow (about 10^{-2} sec) to leave time for the predissociation process to take place.

An example of a different type is provided by the CD_4^+ ion, which above 15.5 eV is observed to split spontaneously into $CD_3^+ + D$ after going through the accelerating field [Dibeler and Rosenstock (29)]. Again there is no metastable excited electronic state of CD_4^+, but the higher vibrational levels of the electronic ground state are sufficiently metastable. The predissociation must be one by vibration, which in a five-atomic molecule may well be slow enough to account for the spontaneous formation of CD_3^+ *after* passage through the accelerating field. The corresponding predissociation process for CH_4^+ has not been observed, presumably because the predissociation probability is much higher, so that decomposition occurs *before* acceleration in the mass spectrometer, and thus CH_3^+ is observed at its normal place. A greater predissociation probability for CH_4^+ than for CD_4^+ has presumably the same reason as for the cases of CH_3, CD_3 and CH_2, CD_2 discussed earlier.

(6) Determination of dissociation energies

An observed predissociation always represents an upper limit to the dissociation energy of the molecule. But even if a sharp predissociation limit is observed, it is not always easy to establish whether or not it corresponds to a precise dissociation limit and whether, therefore, after subtraction of the excitation energy of the products of dissociation, it yields a reliable value for the dissociation energy of the ground state.

A precise determination of the dissociation energy is possible when the beginning of predissociation can be established within the rotational structure for two or more successive vibrational levels. This has been possible for a number of diatomic molecules and free radicals. Figure 106 shows schematically an energy-level diagram in such a case. It can be shown (see *MM* I, p. 430) that, if the breaking-off points in the successive vibrational levels are

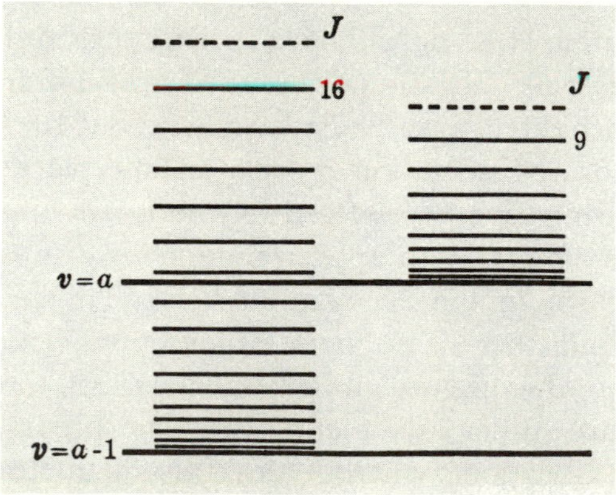

Fig. 106. Breaking-off in two successive vibrational levels; case I of predissociation (from *MM* I, p. 430). The first level that is weakened in emission is indicated by a broken line in each of the vibrational levels.

not very different in energy, the predissociation limit [more accurately obtained by plotting the breaking-off points against $J(J + 1)$ and extrapolating the resulting curve to $J(J + 1) = 0$] is equal to a dissociation limit. In this way very precise dissociation energies of the radicals CH and SO have been obtained, as well as those of many important stable molecules such as N_2 and CO [see *MM* I, p. 442f, and Gaydon (44)].

For polyatomic molecules and radicals there is only one case, that of HNO, where a breaking-off in emission in the rotational structure on account of predissociation has been observed. Although the breaking-off has been observed in two vibrational levels (000 and 010) of the excited electronic state [Clement and Ramsay (19)], it is not possible to draw a conclusion as to whether or not the predissociation limit gives a dissociation limit, since the energy difference of the two breaking-off points is fairly large ($\sim$300 cm^{-1}) and in the opposite direction to that normally observed for diatomic molecules (Fig. 106). The lower breaking-off point does, however, supply a definite upper bound to one of the dissociation energies of HNO—namely, $D_0'' < 2.11$ eV.

For all cases of diffuseness (predissociation) in the absorption spectra of polyatomic molecules only upper limits can be derived for the dissociation energies, and these limits may be quite far above the true values. For example, the predissociation of CH_2, illustrated by Fig. 9, yields an upper bound of 8.8 eV for a dissociation limit. If we assume a fully allowed predissociation, the state causing the predissociation must have the same species as the upper state of the observed bands: $^3\Sigma_u^-$. Such a state cannot arise from CH in the $X^2\Pi$ or $A^2\Delta$ states, but must correspond to $B^2\Sigma^-$. Thus we find the upper limit 5.8 eV for D_0 (CH–H), a value substantially higher than the actual value (estimated from other evidence to be 4.2 eV). The situation in many other cases is similar.

(7) Pre-ionization

If the continuous range of energy levels in Fig. 102(b) corresponds to ionization, the radiationless transition from a discrete state into this continuum leads to ionization of the molecule.

We call this phenomenon *pre-ionization*, in analogy to the term predissociation. However, many authors prefer the name *auto-ionization*.

Pre-ionization may occur for all discrete electronic states that lie above the first ionization potential, subject to selection rules similar to those given above for predissociation. Many instances of pre-ionizations have been observed in the higher Rydberg series of a number of stable molecules. The criterion used in almost all cases was the diffuseness of the corresponding absorption bands. For free radicals, however, no higher Rydberg series have yet been observed and, correspondingly, no cases of pre-ionization.

C. RECOMBINATION

Up to now we have considered processes that lead to dissociation or ionization by light absorption (photodissociation or photoionization). The inverses of these processes are *two-body recombination processes*. The simplest process of this type consists of a direct radiative transition from an upper state belonging to a continuous range to a lower discrete state. Here a continuous emission spectrum arises that is the exact inverse of the continuous absorption spectra considered in section A of this chapter. Since the upper state corresponds to a collision between two atoms or radicals, or between an electron and an ion, and since the collision time is very small (of the order of 10^{-13} sec) compared to the radiative lifetime ($\sim 10^{-8}$ sec), the intensity of such emission continua is very low. It is difficult to establish this mechanism for a given observed emission continuum.

A radiative two-body recombination is also possible by way of an *inverse Auger process* (inverse predissociation or pre-ionization). In this case two particles (radical + radical, radical + atom, atom + atom, or ion + electron) approach one another with an energy equaling that of a "discrete" state of the compound system. There can then be a radiationless transition into this discrete state corresponding to a reversal of the horizontal arrows in Fig. 102(b). After a very short lifetime the radiationless transition will happen again in the opposite direction [corresponding to the arrows in Fig. 102(b)] and the two particles will separate again. However, if during the lifetime of the compound system a radiative transition to a lower stable state takes place, an actual recombination of the atoms or radicals, or ions and electrons, takes place to form a molecule or radical (or ion). This type of recombination process must be clearly distinguished from the direct recombination mentioned in the preceding paragraph.

Whether in a given case the inverse Auger process leads to a greater rate of radiative recombination than the direct process depends on the relative values of the transition probabilities to lower stable states and on the density of predissociating or pre-ionizing states. For electron atomic-ion recombination it has been shown by various authors [Burgess (12), Goldberg (45)] that at higher temperatures the effect of inverse pre-ionization on the rate is very large (increase by a factor of 100 to 1000), a fact that is of considerable importance for an understanding of stellar atmospheres and the solar corona. For radical-atom or radical-radical recombination, to my knowledge, no such calculations have been carried out, but it appears likely that inverse predissociation is an important contributor to the rate of recombination at low pressure, particularly for polyatomic systems. Inverse Auger processes also increase the rate of three-body recombinations, since the lifetime of the collision complex is increased.

We can establish the occurrence of a radiative recombination via an inverse Auger process in a given case either by observing the emission from the diffuse levels or by observing resonances in the scattering cross section of one of the particles against the other. The first method has been used in the detection of inverse predissociation of AlH referred to earlier (see p. 197). It has also been applied to the formation of CO_2 from CO + O, and of NO_2 from NO + O, although in these cases the spectrum of the emission is nearly continuous and does not allow such a definite conclusion as for AlH. The second method has been used in the study of pre-ionized excited states of various negative ions (He^-, H_2^-, N_2^-, and others).

BIBLIOGRAPHY

1. T. Amano, E. Hirota, and Y. Morino, J. Phys. Soc. Japan **22,** 399 (1967).
2. ———, ———, and ———, J. Mol. Spec. **27,** 257 (1968).
3. ———, S. Saito, E. Hirota, Y. Morino, D. R. Johnson, and F. T. Powell, J. Mol. Spec. **30,** 275 (1967).
4. E. A. Ballik and D. A. Ramsay, Astrophys. J. **137,** 84 (1963).
5. J. L. Bancroft, J. M. Hollas, and D. A. Ramsay, Can. J. Phys. **40,** 322 (1962).
6. D. R. Bates and M. Nicolet, Publ. Astron. Soc. Pac. **62,** 106 (1950); J. Geophys. Res. **55,** 301 (1950).
7. S. H. Bauer, G. Herzberg, and J. W. C. Johns, J. Mol. Spec. **13,** 256 (1964).
8. E. Bengtsson-Knave and R. Rydberg, Z. Physik **59,** 540 (1930).
9. H. J. Bernstein and G. Herzberg, J. Chem. Phys. **16,** 30 (1948).
10. K. F. Bonhoeffer and L. Farkas, Z. phys. Chem. A**134,** 337 (1927).
11. ——— and H. Reichardt, Z. phys. Chem. A**139,** 75 (1928).
12. A. Burgess, Astrophys. J. **139,** 776 (1964).
13. G. A. Carlson and G. C. Pimentel, J. Chem. Phys. **44,** 4053 (1966).
14. A. Carrington, Proc. Roy. Soc. **302**A, 291 (1968).

14a. ———, A. R. Fabris, and N. J. D. Lucas, J. Chem. Phys. **49,** 5545 (1968).

15. ——— and D. H. Levy, J. Phys. Chem. **71,** 2 (1967).
16. J. W. Chamberlain and F. L. Roesler, Astrophys. J. **121,** 541 (1955).
17. M. S. Child, J. Mol. Spec. **10,** 357 (1963).
18. ——— and H. C. Longuet-Higgins, Phil. Trans. Roy. Soc. **254**A, 259 (1961).
19. M. J. Y. Clement and D. A. Ramsay, Can. J. Phys. **39,** 205 (1961).
20. R. Colin, Can. J. Phys. **46,** 1539 (1968).
21. ———, Can. J. Phys. **47,** 979 (1969).
22. ——— and W. E. Jones, Can. J. Phys. **45,** 301 (1967).
23. C. A. Coulson, Valence, 2nd ed. (Oxford University Press, New York, 1961).

24. F. Creutzberg, Can. J. Phys. **44,** 1583 (1966).
25. J. Curry, L. Herzberg, and G. Herzberg, Z. Physik **86,** 348 (1933).
26. F. W. Dalby, Can. J. Phys. **36,** 1336 (1958).
27. R. Daudel, Structure Électronique des Molécules (Gauthier-Villars, Paris, 1962).
27a. D. D. Davis and H. Okabe, J. Chem. Phys. **49,** 5526 (1968).
28. C. Devillers and D. A. Ramsay, Can. J. Phys. (to be published).
29. V. H. Dibeler and H. M. Rosenstock, J. Chem. Phys. **39,** 1326 (1963).
30. ——— and ———, J. Chem. Phys. **39,** 3106 (1963).
31. R. N. Dixon, Phil. Trans. Roy. Soc. **252**A, 165 (1960).
32. ———, Trans. Far. Soc. **60,** 1363 (1964).
32a. ———,J. Mol. Phys. (in press).
33. A. E. Douglas, Disc. Far. Soc. **35,** 158 (1963).
34. ——— and G. Herzberg, Can. J. Res. A**18,** 179 (1940).
35. ——— and W. E. Jones, Can. J. Phys. **44,** 2251 (1966).
36. ——— and P. M. Routley, Astrophys. J. **119,** 303 (1954).
37. G. C. Dousmanis, T. M. Sanders, and C. H. Townes, Phys. Rev. **100,** 1735 (1955).
38. K. Dressler and D. A. Ramsay, Phil. Trans. Roy. Soc. **251**A, 553 (1959).
39. P. J. Dyne and D. W. G. Style, Disc. Far. Soc. **2,** 159 (1947).
40. G. Ehrenstein, C. H. Townes, and M. J. Stevenson, Phys. Rev. Lett. **3,** 40 (1959).
41. U. Fano, Phys. Rev. **124,** 1866 (1961).
42. D. Garvin, H. P. Broida, and H. J. Kostkowski, J. Chem. Phys. **32,** 880 (1960).
43. L. Gausset, G. Herzberg, A. Lagerqvist, and B. Rosen, Astrophys. J. **142,** 45 (1965).
44. A. G. Gaydon, Dissociation Energies, 3rd ed. (Chapman and Hall, London, 1968).
45. L. Goldberg, in A. Temkin, ed. Autoionization, p. 1 (Mono Book Corp., Baltimore, 1966).
46. M. Gomberg, Ber. d.d. chem. Ges. **33,** 3150 (1900).
47. W. Gordy, W. V. Smith, and R. F. Trambarulo, Microwave Spectroscopy (Wiley, New York, 1953).
48. H. Hartmann, Theorie der chemischen Bindung auf quantentheoretischer Grundlage (Springer, Berlin, 1954).
49. V. Henri, C.R. (Paris) **177,** 1037 (1923).

49a. ———, Structure des Molécules (Paris 1925).
50. G. Herbig, Publ. Astron. Soc. Pac. **68,** 204 (1956).
51. R. C. Herman and G. A. Hornbeck, Astrophys. J. **118,** 214 (1953).
52. K. C. Herr and G. C. Pimentel, App. Opt. **4,** 25 (1965).
53. G. Herzberg, Rev. Mod. Phys. **14,** 195 (1942).
54. ———, Astrophys. J. **96,** 314 (1942).
55. ———, J. Roy. Astron. Soc. Canada **45,** 100 (1951).
56. ———, Can. J. Phys. **39,** 1511 (1961).
57. ———, Proc. Roy. Soc. **262**A, 291 (1961).
58. ———, J. Opt. Soc. Amer. **55,** 229 (1965).
59. ———, I.A.U. Symp. No. 31, p. 91 (1967).
60. ———, Pont. Acad. Sci. Commentarii **2** No. 15 (1968).
61. ——— (to be published).
62. ——— and L. L. Howe, Can. J. Phys. **37,** 636 (1959).
63. ——— and J. W. C. Johns, Proc. Roy. Soc. **295**A, 107 (1966).
64. ——— and ———, Proc. Roy. Soc. **298**A, 142 (1967).
65. ——— and ———, Astrophys. J. **158,** 399 (1969).
66. ——— and A. Lagerqvist, Can. J. Phys. **46,** 2363 (1968).
67. ———, A. Lagerqvist, and C. Malmberg, Can. J. Phys. **47,** 2735 (1969).
68. ——— and D. A. Ramsay, Proc. Roy. Soc. **233**A, 34 (1955).
69. ——— and E. Teller, Z. phys. Chem. **B21,** 410 (1933).
70. ——— and D. N. Travis, Can. J. Phys. **42,** 1658 (1964).
71. ——— and R. D. Verma, Can. J. Phys. **42,** 395 (1964).
72. ——— and P. A. Warsop, Can. J. Phys. **41,** 286 (1963).
73. J. T. Hougen, J. Chem. Phys. **36,** 519 (1962).
74. ———, J. Chem. Phys. **39,** 358 (1963).
75. H. A. Jahn and E. Teller, Proc. Roy. Soc. **161**A, 220 (1937).
76. K. B. Jefferts, Phys. Rev. Lett. **20,** 39 (1968), **23,** 1476 (1969).
76a. K. B. Jefferts, A. A. Penzias, J. A. Ball, D. F. Dickinson and A. E. Lilley, Astrophys. J. **159,** L15 (1970).
77. F. A. Jenkins, Y. K. Roots, and R. S. Mulliken, Phys. Rev. **39,** 16 (1932).
78. J. W. C. Johns, Can. J. Phys. **39,** 1738 (1961).
79. ———, J. Mol. Spec. **15,** 473 (1965).
80. ———, F. A. Grimm, and R. F. Porter, J. Mol. Spec. **22,** 435 (1967).
81. D. L. Judge *et al.* (to be published).
82. M. Karplus, J. Chem. Phys. **30,** 15 (1959).

83. V. M. Khanna, R. Hauge, R. F. Curl, Jr., and J. L. Margrave, J. Chem. Phys. **47,** 5031 (1967).
84. H. W. Kroto, Can. J. Phys. **45,** 1439 (1967).
85. R. K. Laird, E. B. Andrews, and R. F. Barrow, Trans. Far. Soc. **46,** 803 (1950).
86. H. C. Longuet-Higgins, U. Öpik, M. H. L. Pryce, and R. A. Sack, Proc. Roy. Soc. **244**A, 1 (1958).
87. S. K. Luke, Astrophys. J. **156,** 761 (1969).
87a. R. MacDonald, H. L. Buijs, and H. P. Gush, Can. J. Phys. **46,** 2575 (1968).
88. R. P. Madden and W. S. Benedict, J. Chem. Phys. **23,** 408 (1955).
89. C. W. Mathews, Can. J. Phys. **45,** 2355 (1967).
90. J. D. McKinley, D. Garvin, and M. J. Boudart, J. Chem. Phys. **23,** 784 (1955).
91. A. B. Meinel, Astrophys. J. **111,** 555, **112,** 120 (1950).
92. A. J. Merer and D. N. Travis, Can. J. Phys. **44,** 1541 (1966).
93. F. A. Miller and R. B. Hannan, Jr., Spectrochim. Acta **12,** 321 (1958).
94. D. E. Milligan and M. E. Jacox, J. Chem. Phys. **47,** 5146 (1967).
95. ——— and ———, J. Chem. Phys. **47,** 5157 (1967).
96. ——— and ———, J. Chem. Phys. **51,** 1952 (1969).
97. E. R. V. Milton, N. B. Dunford, and A. E. Douglas, J. Chem. Phys. **35,** 1202 (1961).
98. A. Monfils and B. Rosen, Nature **164,** 713 (1949).
99. R. S. Mulliken, Phys. Rev. **25,** 259 (1925).
100. ——— [unpublished, quoted in (53)].
101. J. N. Murrell, S. F. A. Kettle, and J. M. Tedder, Valence Theory (Wiley, New York, 1965).
102. N. A. Narasimham *et al.* (unpublished).
103. H. Neuimin and A. Terenin, Acta Physicochim. URSS **5,** 465 (1936).
104. R. G. W. Norrish and G. Porter, Nature **164,** 658 (1949).
105. J. F. Ogilvie, Spectrochim. Acta **23**A, 737 (1967).
106. O. Oldenberg, J. Chem. Phys. **2,** 713 (1934).
107. F. Paneth and W. Hofeditz, Ber. d.d. chem. Ges. **62**B, 1335 (1929).
108. R. G. Parr, The Quantum Theory of Molecular Electronic Structure (Benjamin, New York, 1963).
109. L. Pauling, The Nature of the Chemical Bond, 3rd ed. (Cornell University Press, Ithaca, 1960).

110. L. Pauling and E. B. Wilson, Jr., Introduction to Quantum Mechanics (McGraw-Hill, New York, 1935).
111. J. A. Pople, Mol. Phys. **3,** 16 (1960).
112. F. X. Powell and D. R. Lide, Jr., J. Chem. Phys. **41,** 1413 (1964).
113. ——— and ———, J. Chem. Phys. **42,** 4201 (1965).
114. ——— and ———, J. Chem. Phys. **45,** 1067 (1966).
115. H. E. Radford, Phys. Rev. **122,** 114 (1961); J. Chem. Phys. **40,** 2732 (1964).
116. ——— and M. Linzer, Phys. Rev. Lett. **10,** 443 (1963).
117. V. M. Rao and R. F. Curl, Jr., J. Chem. Phys. **45,** 2032 (1966).
118. ———, ———, P. L. Timms, and J. L. Margrave, J. Chem. Phys. **43,** 2557 (1965).
119. W. T. Raynes, J. Chem. Phys. **41,** 3020 (1964).
120. R. Renner, Z. Physik **92,** 172 (1934).
121. S. Saito and T. Amano, J. Mol. Spec. **34,** 383 (1970)
122. W. B. Somerville, Monthly Notices, R.A.S. **147,** 201 (1970).
123. G. Stenvinkel, Z. Physik **114,** 602 (1939).
124. B. P. Stoicheff, Advances in Spectroscopy **1,** 91 (1959).
125. A. Streitwieser, Jr., Molecular Orbital Theory for Organic Chemists (Wiley, New York, 1961).
126. D. W. G. Style and J. C. Ward, J. Chem. Soc. p. 2125 (1952); Trans. Far. Soc. **49,** 999 (1953).
127. T. M. Sugden and C. N. Kenney, Microwave Spectroscopy of Gases (Van Nostrand, London, 1965).
127a. P. Swings, Publ. Astron. Soc. Pac. **54,** 123 (1942).
128. P. Swings, C. T. Elvey, and H. W. Babcock, Astrophys. J. **94,** 320 (1941).
129. H. S. Taylor, Trans. Far. Soc. **21,** 560 (1925).
130. A. N. Terenin, Usp. Fiz. Nauk **36,** 292 (1948).
131. W. R. Thorson and I. Nakagawa, J. Chem. Phys. **33,** 994 (1960).
132. C. H. Townes and A. L. Schawlow, Microwave Spectroscopy (McGraw-Hill, New York, 1955).
133. A. Vallance Jones, Mem. Soc. Roy. Sci. Liège **9,** 289 (1964).
134. R. D. Verma and P. A. Warsop, Can. J. Phys. **41,** 152 (1963).
135. A. D. Walsh, J. Chem. Soc. pp. 2260, 2266, 2288, 2296 (1953).
136. S. Weissman, J. T. Vanderslice, and R. Battino, J. Chem. Phys. **39,** 2226 (1963).
137. W. Weizel, Z. Physik **54,** 321 (1929).
138. J. U. White, J. Opt. Soc. Amer. **32,** 285 (1942).

139. G. Winnewisser, M. Winnewisser, and W. Gordy, J. Chem. Phys. **49,** 3465 (1968).
140. M. Winnewisser, K. V. L. N. Sastry, R. L. Cook, and W. Gordy, J. Chem. Phys. **41,** 1687 (1964).
141. J. E. Wollrab, Rotational Spectra and Molecular Structure (Academic Press, New York, 1967).
142. R. W. Wood, Phil. Mag. **42,** 729 (1921); Proc. Roy. Soc. **102,** 1 (1922).
143. A. R. W. McKellar, P. R. Bunker, T. J. Sears, K. M. Evenson, R. J. Saykally, and S. R. Langhoff, J. Chem. Phys. **79**, 5251 (1983).
144. P. R. Bunker, P. Jensen, W. P. Kraemer, and R. Beardsworth, J. Chem. Phys. **85**, 3724 (1986).
145. M. Elitzur, Astrophys. J. **218**, 677 (1977).

INDEX

This index includes all symbols used and all molecules and radicals discussed or mentioned. The symbols are listed at the beginning of the section devoted to the particular letter. Greek letter symbols are given under the letter with which they begin when they are written in English (for example, ϕ, π, ψ, under P and in this order). Symbols to which a word is joined are listed under the corresponding symbol: for example, "*R* branch" is under R, not under Rb. In all other cases the alphabeting is based on the part before the comma; for example, "electron spin" comes after "electronic angular momentum."

Individual radicals or molecules are listed under their usual chemical formulae considered as words; for example, CH_3 under Ch. If there are several radicals giving the same "word" they are listed in order of increasing numbers of the first, second, . . . atom; for example, CH, CH_2, CH_3, CH_4^+ in this order.

Italicized page numbers refer to definitions or more detailed discussions, ordinary (roman) page numbers to brief or casual references to the item listed, boldface page numbers refer to illustrations.

Index

Index

Index

Index

A CATALOG OF SELECTED

DOVER BOOKS

IN ALL FIELDS OF INTEREST

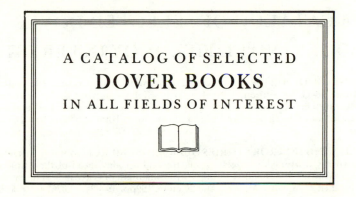

A CATALOG OF SELECTED DOVER BOOKS IN ALL FIELDS OF INTEREST

DRAWINGS OF REMBRANDT, edited by Seymour Slive. Updated Lippmann, Hofstede de Groot edition, with definitive scholarly apparatus. All portraits, biblical sketches, landscapes, nudes. Oriental figures, classical studies, together with selection of work by followers. 550 illustrations. Total of 630pp. 9⅛ × 12¼.
21485-0, 21486-9 Pa., Two-vol. set $25.00

GHOST AND HORROR STORIES OF AMBROSE BIERCE, Ambrose Bierce. 24 tales vividly imagined, strangely prophetic, and decades ahead of their time in technical skill: "The Damned Thing," "An Inhabitant of Carcosa," "The Eyes of the Panther," "Moxon's Master," and 20 more. 199pp. 5⅜ × 8½. 20767-6 Pa. $3.95

ETHICAL WRITINGS OF MAIMONIDES, Maimonides. Most significant ethical works of great medieval sage, newly translated for utmost precision, readability. Laws Concerning Character Traits, Eight Chapters, more. 192pp. 5⅜ × 8½.
24522-5 Pa. $4.50

THE EXPLORATION OF THE COLORADO RIVER AND ITS CANYONS, J. W. Powell. Full text of Powell's 1,000-mile expedition down the fabled Colorado in 1869. Superb account of terrain, geology, vegetation, Indians, famine, mutiny, treacherous rapids, mighty canyons, during exploration of last unknown part of continental U.S. 400pp. 5⅜ × 8½. 20094-9 Pa. $6.95

HISTORY OF PHILOSOPHY, Julián Marías. Clearest one-volume history on the market. Every major philosopher and dozens of others, to Existentialism and later. 505pp. 5⅜ × 8½. 21739-6 Pa. $8.50

ALL ABOUT LIGHTNING, Martin A. Uman. Highly readable non-technical survey of nature and causes of lightning, thunderstorms, ball lightning, St. Elmo's Fire, much more. Illustrated. 192pp. 5⅜ × 8½. 25237-X Pa. $5.95

SAILING ALONE AROUND THE WORLD, Captain Joshua Slocum. First man to sail around the world, alone, in small boat. One of great feats of seamanship told in delightful manner. 67 illustrations. 294pp. 5⅜ × 8½. 20326-3 Pa. $4.95

LETTERS AND NOTES ON THE MANNERS, CUSTOMS AND CONDITIONS OF THE NORTH AMERICAN INDIANS, George Catlin. Classic account of life among Plains Indians: ceremonies, hunt, warfare, etc. 312 plates. 572pp. of text. 6⅛ × 9¼. 22118-0, 22119-9 Pa. Two-vol. set $15.90

ALASKA: The Harriman Expedition, 1899, John Burroughs, John Muir, et al. Informative, engrossing accounts of two-month, 9,000-mile expedition. Native peoples, wildlife, forests, geography, salmon industry, glaciers, more. Profusely illustrated. 240 black-and-white line drawings. 124 black-and-white photographs. 3 maps. Index. 576pp. 5⅜ × 8½. 25109-8 Pa. $11.95

HOW TO WRITE, Gertrude Stein. Gertrude Stein claimed anyone could understand her unconventional writing—here are clues to help. Fascinating improvisations, language experiments, explanations illuminate Stein's craft and the art of writing. Total of 414pp. 4⅜ × 6⅜. 23144-5 Pa. $5.95

ADVENTURES AT SEA IN THE GREAT AGE OF SAIL: Five Firsthand Narratives, edited by Elliot Snow. Rare true accounts of exploration, whaling, shipwreck, fierce natives, trade, shipboard life, more. 33 illustrations. Introduction. 353pp. 5⅜ × 8½. 25177-2 Pa. $7.95

THE HERBAL OR GENERAL HISTORY OF PLANTS, John Gerard. Classic descriptions of about 2,850 plants—with over 2,700 illustrations—includes Latin and English names, physical descriptions, varieties, time and place of growth, more. 2,706 illustrations. xlv + 1,678pp. 8½ × 12¼. 23147-X Cloth. $75.00

DOROTHY AND THE WIZARD IN OZ, L. Frank Baum. Dorothy and the Wizard visit the center of the Earth, where people are vegetables, glass houses grow and Oz characters reappear. Classic sequel to *Wizard of Oz*. 256pp. 5⅜ × 8. 24714-7 Pa. $4.95

SONGS OF EXPERIENCE: Facsimile Reproduction with 26 Plates in Full Color, William Blake. This facsimile of Blake's original "Illuminated Book" reproduces 26 full-color plates from a rare 1826 edition. Includes "The Tyger," "London," "Holy Thursday," and other immortal poems. 26 color plates. Printed text of poems. 48pp. 5¼ × 7. 24636-1 Pa. $3.50

SONGS OF INNOCENCE, William Blake. The first and most popular of Blake's famous "Illuminated Books," in a facsimile edition reproducing all 31 brightly colored plates. Additional printed text of each poem. 64pp. 5¼ × 7. 22764-2 Pa. $3.50

PRECIOUS STONES, Max Bauer. Classic, thorough study of diamonds, rubies, emeralds, garnets, etc.: physical character, occurrence, properties, use, similar topics. 20 plates, 8 in color. 94 figures. 659pp. 6⅛ × 9¼. 21910-0, 21911-9 Pa., Two-vol. set $15.90

ENCYCLOPEDIA OF VICTORIAN NEEDLEWORK, S. F. A. Caulfeild and Blanche Saward. Full, precise descriptions of stitches, techniques for dozens of needlecrafts—most exhaustive reference of its kind. Over 800 figures. Total of 679pp. 8⅛ × 11. Two volumes. Vol. 1 22800-2 Pa. $11.95
Vol. 2 22801-0 Pa. $11.95

THE MARVELOUS LAND OF OZ, L. Frank Baum. Second Oz book, the Scarecrow and Tin Woodman are back with hero named Tip, Oz magic. 136 illustrations. 287pp. 5⅜ × 8½. 20692-0 Pa. $5.95

WILD FOWL DECOYS, Joel Barber. Basic book on the subject, by foremost authority and collector. Reveals history of decoy making and rigging, place in American culture, different kinds of decoys, how to make them, and how to use them. 140 plates. 156pp. 7⅞ × 10¾. 20011-6 Pa. $8.95

HISTORY OF LACE, Mrs. Bury Palliser. Definitive, profusely illustrated chronicle of lace from earliest times to late 19th century. Laces of Italy, Greece, England, France, Belgium, etc. Landmark of needlework scholarship. 266 illustrations. 672pp. 6⅛ × 9¼. 24742-2 Pa. $14.95

ILLUSTRATED GUIDE TO SHAKER FURNITURE, Robert Meader. All furniture and appurtenances, with much on unknown local styles. 235 photos. 146pp. 9 × 12. 22819-3 Pa. $7.95

WHALE SHIPS AND WHALING: A Pictorial Survey, George Francis Dow. Over 200 vintage engravings, drawings, photographs of barks, brigs, cutters, other vessels. Also harpoons, lances, whaling guns, many other artifacts. Comprehensive text by foremost authority. 207 black-and-white illustrations. 288pp. 6 × 9. 24808-9 Pa. $8.95

THE BERTRAMS, Anthony Trollope. Powerful portrayal of blind self-will and thwarted ambition includes one of Trollope's most heartrending love stories. 497pp. 5⅜ × 8½. 25119-5 Pa. $8.95

ADVENTURES WITH A HAND LENS, Richard Headstrom. Clearly written guide to observing and studying flowers and grasses, fish scales, moth and insect wings, egg cases, buds, feathers, seeds, leaf scars, moss, molds, ferns, common crystals, etc.—all with an ordinary, inexpensive magnifying glass. 209 exact line drawings aid in your discoveries. 220pp. 5⅜ × 8½. 23330-8 Pa. $4.50

RODIN ON ART AND ARTISTS, Auguste Rodin. Great sculptor's candid, wide-ranging comments on meaning of art; great artists; relation of sculpture to poetry, painting, music; philosophy of life, more. 76 superb black-and-white illustrations of Rodin's sculpture, drawings and prints. 119pp. 8⅜ × 11¼. 24487-3 Pa. $6.95

FIFTY CLASSIC FRENCH FILMS, 1912–1982: A Pictorial Record, Anthony Slide. Memorable stills from Grand Illusion, Beauty and the Beast, Hiroshima, Mon Amour, many more. Credits, plot synopses, reviews, etc. 160pp. 8¼ × 11. 25256-6 Pa. $11.95

THE PRINCIPLES OF PSYCHOLOGY, William James. Famous long course complete, unabridged. Stream of thought, time perception, memory, experimental methods; great work decades ahead of its time. 94 figures. 1,391pp. 5⅜ × 8½. 20381-6, 20382-4 Pa., Two-vol. set $19.90

BODIES IN A BOOKSHOP, R. T. Campbell. Challenging mystery of blackmail and murder with ingenious plot and superbly drawn characters. In the best tradition of British suspense fiction. 192pp. 5⅜ × 8½. 24720-1 Pa. $3.95

CALLAS: PORTRAIT OF A PRIMA DONNA, George Jellinek. Renowned commentator on the musical scene chronicles incredible career and life of the most controversial, fascinating, influential operatic personality of our time. 64 black-and-white photographs. 416pp. 5⅜ × 8¼. 25047-4 Pa. $7.95

GEOMETRY, RELATIVITY AND THE FOURTH DIMENSION, Rudolph Rucker. Exposition of fourth dimension, concepts of relativity as Flatland characters continue adventures. Popular, easily followed yet accurate, profound. 141 illustrations. 133pp. 5⅜ × 8½. 23400-2 Pa. $3.50

HOUSEHOLD STORIES BY THE BROTHERS GRIMM, with pictures by Walter Crane. 53 classic stories—Rumpelstiltskin, Rapunzel, Hansel and Gretel, the Fisherman and his Wife, Snow White, Tom Thumb, Sleeping Beauty, Cinderella, and so much more—lavishly illustrated with original 19th century drawings. 114 illustrations. x + 269pp. 5⅜ × 8½. 21080-4 Pa. $4.50

SUNDIALS, Albert Waugh. Far and away the best, most thorough coverage of ideas, mathematics concerned, types, construction, adjusting anywhere. Over 100 illustrations. 230pp. 5⅜ × 8½. 22947-5 Pa. $4.50

PICTURE HISTORY OF THE NORMANDIE: With 190 Illustrations, Frank O. Braynard. Full story of legendary French ocean liner: Art Deco interiors, design innovations, furnishings, celebrities, maiden voyage, tragic fire, much more. Extensive text. 144pp. 8⅞ × 11¾. 25257-4 Pa. $9.95

THE FIRST AMERICAN COOKBOOK: A Facsimile of "American Cookery," 1796, Amelia Simmons. Facsimile of the first American-written cookbook published in the United States contains authentic recipes for colonial favorites—pumpkin pudding, winter squash pudding, spruce beer, Indian slapjacks, and more. Introductory Essay and Glossary of colonial cooking terms. 80pp. 5⅜ × 8½. 24710-4 Pa. $3.50

101 PUZZLES IN THOUGHT AND LOGIC, C. R. Wylie, Jr. Solve murders and robberies, find out which fishermen are liars, how a blind man could possibly identify a color—purely by your own reasoning! 107pp. 5⅜ × 8½. 20367-0 Pa. $2.50

THE BOOK OF WORLD-FAMOUS MUSIC—CLASSICAL, POPULAR AND FOLK, James J. Fuld. Revised and enlarged republication of landmark work in musico-bibliography. Full information about nearly 1,000 songs and compositions including first lines of music and lyrics. New supplement. Index. 800pp. 5⅜ × 8¼. 24857-7 Pa. $14.95

ANTHROPOLOGY AND MODERN LIFE, Franz Boas. Great anthropologist's classic treatise on race and culture. Introduction by Ruth Bunzel. Only inexpensive paperback edition. 255pp. 5⅜ × 8½. 25245-0 Pa. $5.95

THE TALE OF PETER RABBIT, Beatrix Potter. The inimitable Peter's terrifying adventure in Mr. McGregor's garden, with all 27 wonderful, full-color Potter illustrations. 55pp. 4¼ × 5½. (Available in U.S. only) 22827-4 Pa. $1.75

THREE PROPHETIC SCIENCE FICTION NOVELS, H. G. Wells. *When the Sleeper Wakes, A Story of the Days to Come* and *The Time Machine* (full version). 335pp. 5⅜ × 8½. (Available in U.S. only) 20605-X Pa. $5.95

APICIUS COOKERY AND DINING IN IMPERIAL ROME, edited and translated by Joseph Dommers Vehling. Oldest known cookbook in existence offers readers a clear picture of what foods Romans ate, how they prepared them, etc. 49 illustrations. 301pp. 6⅛ × 9¼. 23563-7 Pa. $6.50

SHAKESPEARE LEXICON AND QUOTATION DICTIONARY, Alexander Schmidt. Full definitions, locations, shades of meaning of every word in plays and poems. More than 50,000 exact quotations. 1,485pp. 6½ × 9¼. 22726-X, 22727-8 Pa., Two-vol. set $27.90

THE WORLD'S GREAT SPEECHES, edited by Lewis Copeland and Lawrence W. Lamm. Vast collection of 278 speeches from Greeks to 1970. Powerful and effective models; unique look at history. 842pp. 5⅜ × 8½. 20468-5 Pa. $11.95

THE ART NOUVEAU STYLE BOOK OF ALPHONSE MUCHA: All 72 Plates from "Documents Decoratifs" in Original Color, Alphonse Mucha. Rare copyright-free design portfolio by high priest of Art Nouveau. Jewelry, wallpaper, stained glass, furniture, figure studies, plant and animal motifs, etc. Only complete one-volume edition. 80pp. 9⅜ × 12¼. 24044-4 Pa. $8.95

ANIMALS: 1,419 COPYRIGHT-FREE ILLUSTRATIONS OF MAMMALS, BIRDS, FISH, INSECTS, ETC., edited by Jim Harter. Clear wood engravings present, in extremely lifelike poses, over 1,000 species of animals. One of the most extensive pictorial sourcebooks of its kind. Captions. Index. 284pp. 9 × 12. 23766-4 Pa. $9.95

OBELISTS FLY HIGH, C. Daly King. Masterpiece of American detective fiction, long out of print, involves murder on a 1935 transcontinental flight—"a very thrilling story"—NY Times. Unabridged and unaltered republication of the edition published by William Collins Sons & Co. Ltd., London, 1935. 288pp. 5⅜ × 8½. (Available in U.S. only) 25036-9 Pa. $4.95

VICTORIAN AND EDWARDIAN FASHION: A Photographic Survey, Alison Gernsheim. First fashion history completely illustrated by contemporary photographs. Full text plus 235 photos, 1840–1914, in which many celebrities appear. 240pp. 6½ × 9¼. 24205-6 Pa. $6.00

THE ART OF THE FRENCH ILLUSTRATED BOOK, 1700–1914, Gordon N. Ray. Over 630 superb book illustrations by Fragonard, Delacroix, Daumier, Doré, Grandville, Manet, Mucha, Steinlen, Toulouse-Lautrec and many others. Preface. Introduction. 633 halftones. Indices of artists, authors & titles, binders and provenances. Appendices. Bibliography. 608pp. 8⅛ × 11¼. 25086-5 Pa. $24.95

THE WONDERFUL WIZARD OF OZ, L. Frank Baum. Facsimile in full color of America's finest children's classic. 143 illustrations by W. W. Denslow. 267pp. 5⅜ × 8½. 20691-2 Pa. $5.95

FRONTIERS OF MODERN PHYSICS: New Perspectives on Cosmology, Relativity, Black Holes and Extraterrestrial Intelligence, Tony Rothman, et al. For the intelligent layman. Subjects include: cosmological models of the universe; black holes; the neutrino; the search for extraterrestrial intelligence. Introduction. 46 black-and-white illustrations. 192pp. 5⅜ × 8½. 24587-X Pa. $6.95

THE FRIENDLY STARS, Martha Evans Martin & Donald Howard Menzel. Classic text marshalls the stars together in an engaging, non-technical survey, presenting them as sources of beauty in night sky. 23 illustrations. Foreword. 2 star charts. Index. 147pp. 5⅜ × 8½. 21099-5 Pa. $3.50

FADS AND FALLACIES IN THE NAME OF SCIENCE, Martin Gardner. Fair, witty appraisal of cranks, quacks, and quackeries of science and pseudoscience: hollow earth, Velikovsky, orgone energy, Dianetics, flying saucers, Bridey Murphy, food and medical fads, etc. Revised, expanded In the Name of Science. "A very able and even-tempered presentation."—The New Yorker. 363pp. 5⅜ × 8. 20394-8 Pa. $6.50

ANCIENT EGYPT: ITS CULTURE AND HISTORY, J. E Manchip White. From pre-dynastics through Ptolemies: society, history, political structure, religion, daily life, literature, cultural heritage. 48 plates. 217pp. 5⅜ × 8½. 22548-8 Pa. $4.95

PLANTS OF THE BIBLE, Harold N. Moldenke and Alma L. Moldenke. Standard reference to all 230 plants mentioned in Scriptures. Latin name, biblical reference, uses, modern identity, much more. Unsurpassed encyclopedic resource for scholars, botanists, nature lovers, students of Bible. Bibliography. Indexes. 123 black-and-white illustrations. 384pp. 6 × 9. 25069-5 Pa. $8.95

FAMOUS AMERICAN WOMEN: A Biographical Dictionary from Colonial Times to the Present, Robert McHenry, ed. From Pocahontas to Rosa Parks, 1,035 distinguished American women documented in separate biographical entries. Accurate, up-to-date data, numerous categories, spans 400 years. Indices. 493pp. 6½ × 9¼. 24523-3 Pa. $9.95

THE FABULOUS INTERIORS OF THE GREAT OCEAN LINERS IN HISTORIC PHOTOGRAPHS, William H. Miller, Jr. Some 200 superb photographs capture exquisite interiors of world's great "floating palaces"—1890's to 1980's: *Titanic, Ile de France, Queen Elizabeth, United States, Europa,* more. Approx. 200 black-and-white photographs. Captions. Text. Introduction. 160pp. 8⅜ × 11¾. 24756-2 Pa. $9.95

THE GREAT LUXURY LINERS, 1927–1954: A Photographic Record, William H. Miller, Jr. Nostalgic tribute to heyday of ocean liners. 186 photos of Ile de France, Normandie, Leviathan, Queen Elizabeth, United States, many others. Interior and exterior views. Introduction. Captions. 160pp. 9 × 12. 24056-8 Pa. $9.95

A NATURAL HISTORY OF THE DUCKS, John Charles Phillips. Great landmark of ornithology offers complete detailed coverage of nearly 200 species and subspecies of ducks: gadwall, sheldrake, merganser, pintail, many more. 74 full-color plates, 102 black-and-white. Bibliography. Total of 1,920pp. 8⅜ × 11¼. 25141-1, 25142-X Cloth. Two-vol. set $100.00

THE SEAWEED HANDBOOK: An Illustrated Guide to Seaweeds from North Carolina to Canada, Thomas F. Lee. Concise reference covers 78 species. Scientific and common names, habitat, distribution, more. Finding keys for easy identification. 224pp. 5⅜ × 8½. 25215-9 Pa. $5.95

THE TEN BOOKS OF ARCHITECTURE: The 1755 Leoni Edition, Leon Battista Alberti. Rare classic helped introduce the glories of ancient architecture to the Renaissance. 68 black-and-white plates. 336pp. 8⅜ × 11¼. 25239-6 Pa. $14.95

MISS MACKENZIE, Anthony Trollope. Minor masterpieces by Victorian master unmasks many truths about life in 19th-century England. First inexpensive edition in years. 392pp. 5⅜ × 8½. 25201-9 Pa. $7.95

THE RIME OF THE ANCIENT MARINER, Gustave Doré, Samuel Taylor Coleridge. Dramatic engravings considered by many to be his greatest work. The terrifying space of the open sea, the storms and whirlpools of an unknown ocean, the ice of Antarctica, more—all rendered in a powerful, chilling manner. Full text. 38 plates. 77pp. 9¼ × 12. 22305-1 Pa. $4.95

THE EXPEDITIONS OF ZEBULON MONTGOMERY PIKE, Zebulon Montgomery Pike. Fascinating first-hand accounts (1805–6) of exploration of Mississippi River, Indian wars, capture by Spanish dragoons, much more. 1,088pp. 5⅜ × 8½. 25254-X, 25255-8 Pa. Two-vol. set $23.90

SIR HARRY HOTSPUR OF HUMBLETHWAITE, Anthony Trollope. Incisive, unconventional psychological study of a conflict between a wealthy baronet, his idealistic daughter, and their scapegrace cousin. The 1870 novel in its first inexpensive edition in years. 250pp. 5⅜ × 8½. 24953-0 Pa. $5.95

LASERS AND HOLOGRAPHY, Winston E. Kock. Sound introduction to burgeoning field, expanded (1981) for second edition. Wave patterns, coherence, lasers, diffraction, zone plates, properties of holograms, recent advances. 84 illustrations. 160pp. 5⅜ × 8¼. (Except in United Kingdom) 24041-X Pa. $3.50

INTRODUCTION TO ARTIFICIAL INTELLIGENCE: SECOND, ENLARGED EDITION, Philip C. Jackson, Jr. Comprehensive survey of artificial intelligence—the study of how machines (computers) can be made to act intelligently. Includes introductory and advanced material. Extensive notes updating the main text. 132 black-and-white illustrations. 512pp. 5⅜ × 8½. 24864-X Pa. $8.95

HISTORY OF INDIAN AND INDONESIAN ART, Ananda K. Coomaraswamy. Over 400 illustrations illuminate classic study of Indian art from earliest Harappa finds to early 20th century. Provides philosophical, religious and social insights. 304pp. 6⅝ × 9⅝. 25005-9 Pa. $8.95

THE GOLEM, Gustav Meyrink. Most famous supernatural novel in modern European literature, set in Ghetto of Old Prague around 1890. Compelling story of mystical experiences, strange transformations, profound terror. 13 black-and-white illustrations. 224pp. 5⅜ × 8½. (Available in U.S. only) 25025-3 Pa. $5.95

ARMADALE, Wilkie Collins. Third great mystery novel by the author of *The Woman in White* and *The Moonstone*. Original magazine version with 40 illustrations. 597pp. 5⅜ × 8½. 23429-0 Pa. $9.95

PICTORIAL ENCYCLOPEDIA OF HISTORIC ARCHITECTURAL PLANS, DETAILS AND ELEMENTS: With 1,880 Line Drawings of Arches, Domes, Doorways, Facades, Gables, Windows, etc., John Theodore Haneman. Sourcebook of inspiration for architects, designers, others. Bibliography. Captions. 141pp. 9 × 12. 24605-1 Pa. $6.95

BENCHLEY LOST AND FOUND, Robert Benchley. Finest humor from early 30's, about pet peeves, child psychologists, post office and others. Mostly unavailable elsewhere. 73 illustrations by Peter Arno and others. 183pp. 5⅜ × 8½. 22410-4 Pa. $3.95

ERTÉ GRAPHICS, Erté. Collection of striking color graphics: *Seasons, Alphabet, Numerals, Aces* and *Precious Stones*. 50 plates, including 4 on covers. 48pp. 9⅜ × 12¼. 23580-7 Pa. $6.95

THE JOURNAL OF HENRY D. THOREAU, edited by Bradford Torrey, F. H. Allen. Complete reprinting of 14 volumes, 1837-61, over two million words; the sourcebooks for *Walden*, etc. Definitive. All original sketches, plus 75 photographs. 1,804pp. 8½ × 12¼. 20312-3, 20313-1 Cloth., Two-vol. set $80.00

CASTLES: THEIR CONSTRUCTION AND HISTORY, Sidney Toy. Traces castle development from ancient roots. Nearly 200 photographs and drawings illustrate moats, keeps, baileys, many other features. Caernarvon, Dover Castles, Hadrian's Wall, Tower of London, dozens more. 256pp. 5⅜ × 8¼. 24898-4 Pa. $5.95